6th International Conference on Mechanics and Mechatronics (ICMMR 2019)

IOP Conference Series: Materials Science and Engineering
Volume 626

Chongqing, China
18 – 21 July 2019

ISBN: 978-1-5108-9830-1
ISSN: 1757-8981

Printed from e-media with permission by:

Curran Associates, Inc.
57 Morehouse Lane
Red Hook, NY 12571

Some format issues inherent in the e-media version may also appear in this print version.

Institute of Physics
Dirac House, Temple Back
Bristol BS1 6BE UK

Phone: 44 1 17 929 7481
Fax: 44 1 17 920 0979

techtracking@iop.org

Additional copies of this publication are available from:

Curran Associates, Inc.
57 Morehouse Lane
Red Hook, NY 12571 USA
Phone: 845-758-0400
Fax: 845-758-2633
Email: curran@proceedings.com
Web: www.proceedings.com

Table of contents

Volume 626

6th International Conference on Mechanics and Mechatronics Research (ICMMR 2019)

18–21 July 2019, Chongqing, China

Accepted papers received: 27 August 2019
Published online: 2 October 2019

Preface

Papers

Mechatronics Engineering

Mechanical Engineering

IOP Conf. Series: Materials Science and Engineering **626** (2019) 011001 doi:10.1088/1757-899X/626/1/011001

PREFACE

It is with great pleasure that I introduce the proceedings of the 6th International Conference on Mechanics and Mechatronics Research (ICMMR 2019) which was held recently in Chongqing, China during July 18-21. ICMMR 2019 was dedicated to technical issues related to Mechanics and Mechatronics Research.

The conference was held in Chongqing University this year, Chongqing University (CQU) is one of the top 1% Chinese universities within the framework of the State 211 Project and 985 Project. The conference impressed the attendees a lot with the generous help of CQU, further celebrating CQU's 90 anniversaries.

The objectives of the conference were not only to provide opportunities to the international scientists to present state of the art research in Mechanics and Mechatronics Research but also to provide a forum to discuss ideas for future directions in the field and explore possibilities for collaborative research projects.

The conference program included keynote, oral, and poster presentations from scholars working in the areas of Mechanics and Mechatronics Research from all over the world. It covered recent trends and progress made in the field of Mechanics and Mechatronics Research. Keynote lectures regarding the latest information in their respective areas of expertise were shared by experts.

The proceedings present a selection of high quality papers submitted to the conference by researchers from universities, research institutes, and industry. All papers were subjected to peer-review by conference committee members and international reviewers. The papers were selected based on their quality and their relevance to the conference. The proceedings present recent advances in the fields of Mechatronics Engineering as well as Mechanical Engineering.

I would like to express special gratitude to members of the conference committee and organizers of the conference. I would also like to thank the reviewers for their valuable time and advice which helped in improving the quality of the papers selected for presentation at the conference and for publication in the proceedings. Finally I want to thank the authors, the members of the organizing committee, the reviewers, the chairpersons, sponsors, and all other conference participants for their support of ICMMR 2019.

Prof. Lichun Dong and Prof. Baoshan Hu
Chongqing University, China

IOP Conf. Series: Materials Science and Engineering **626** (2019) 011002 doi:10.1088/1757-899X/626/1/011002

Conference Committee Chairs

Lichun Dong, Chongqing University, China
Ramesh K. Agarwal, Washington University in St. Louis, USA

Steering Chair

Baoshan Hu, Chongqing University, China

Steering Co-Chair

Xiaohong Zhu, Sichuan University, China

Program Committee Chairs

Shih-Chieh Lin, National Tsing-Hua University, Taiwan

International Technical Committees

Ch. Sanjay, GITAM University Hyderabad, India
Doty Dewi Risanti, Institut Teknologi Sepuluh Nopember Surabaya,Indonesia
Erna Astuti, Universitas Ahmad Dahlan, Indonesia
Hamdan Haji Ya, Universiti Teknologi PETRONAS (UTP), Malaysia
Iis Sopyan, International Islamic University Malaysia, Malaysia
Kaykhosrow Khojier, Islamic Azad University, Iran
Kevin Amith Mathias, Manipal Academy of Higher Education, India
Megat Ahmad Kamal Megat Hanafiah, Universitit Teknologi MARA Cawangan Pahang, Malaysia
Nasreen Bano, Szabist, Karachi Campus, Pakistan
Nikolay Khokhlov, Southwest State University, Russia
Reza Rashidi, State University of New York, Alfred State College, USA
S.M Sapuan, Universiti Putra malaysia, Malaysia
Sammy V. Militante, University of Antique, Philippines
Saravanan Pichiah, Indian Institute of Technology, India
Shu-San Gan, Petra Christian University, Indonesia
Ying-Hao Chu, National Chiao Tung University, Hsinchu, Taiwan
Yong SUN, Commonwealth Scientific and Industrial Research Organization, Australia

IOP Conf. Series: Materials Science and Engineering **626** (2019) 011003 doi:10.1088/1757-899X/626/1/011003

Peer review statement

All papers published in this volume of *IOP Conference Series: Materials Science and Engineering* have been peer reviewed through processes administered by the proceedings Editors. Reviews were conducted by expert referees to the professional and scientific standards expected of a proceedings journal published by IOP Publishing.

IOP Conf. Series: Materials Science and Engineering **626** (2019) 012001 doi:10.1088/1757-899X/626/1/012001

Organizational Level Cloud Diagnostics System of Complex Electromechanical System Based on SOA

Deyong Cai[1], Yingying Zhao[2], Jing Deng[1] and Dongchen Zhang[1]

1. 32286 Unit of PLA, Ji'nan, China;
2. Shandong Urban Construction Vocational College, Ji'nan, China

deyongcai@163.com

Abstract. In order to better solve the problem of insufficient support ability of high-tech electromechanical system, the cloud computing model was introduced into fault diagnostics system of electromechanical system. An open, loose-coupled architecture of Integrated Diagnostic System (IDS) for electromechanical system is put forward basing on Service-Oriented Cloud Computing Architecture (SOCCA). Test and diagnosis resources of repair shop at all levels and total life cycle information resources of electromechanical system were integrated by Portable Maintenance Aid (PMA) with cloud diagnostics platform. According to different demands, Test and diagnosis resources can be divided into data, software, platform and infrastructure, which can be provided on demand to maintainer of all levels as a service. This allows more complex maintenance issues to be accomplished at Forward Operating Locations (FOLs) by better utilizing the most relevant diagnostic information to enable direct, real-time, and engineering assistance when needed. Organizational Level (O-Level) IDS can effectively improve O-level maintainer support ability of complex electromechanical system.

1. Introduction

With the development in technologies, the performance of electromechanical system has become more and more advanced. At the same time, modern electromechanical system often include a large number of working units or sub-components, which are subject to failures or malfunctions during operations. The complexities and interactions of these components lead to increased difficulties in ensuring the safe and reliable operation of the systems[1,2]. The diagnostic complexity of critical electromechanical system, coupled with the on-going need to support the repair and diagnosis at O-Level yields a demanding set of requirements for O-Level IDS. The Integrated Diagnosis of the electromechanical system is one kind of configurable design and management process that it considers all related diagnosis factor (such as testability, diagnosis, test software and hardware, technical data, training and so on), so as to obtain the best efficiency diagnosis potency[3,4].The proposed O-Level IDS that has the real-time, remote diagnosis capability can improve maintenance practices and support of the O-Level maintainer[5]. The O-Level IDS employs a combination of PMA with net-centric maintenance paradigm to provide the ability for an O-level maintainer to have access to expert knowledge and needed information on-platform[6]. This allows more complex on-platform maintenance issues to be accomplished at FOLs by better utilizing the most relevant diagnostic information to enable direct, real-time, and engineering assistance when needed. In recent years, the developments and achievements in the field of the model-based diagnostics[7], human-machine interfaces[8], electronics and embedded system technologies[9] have investigated the feasibility and commenced development for an innovative and advanced O-Level IDS.

This paper is organized as follows: Section II, basing on Service Oriented Cloud Computing Architecture, test and diagnosis resources are divided into four categories, an open, loose-coupled Architecture of O-Level Integrated Diagnostics is put forward. In Section III, the model of O-Level IDS for electromechanical system is presented. Finally, some concluding remarks are made.

2. O-Level Integrated Diagnostics Architecture

2.1. Service Oriented Cloud Computing Architecture

Service-Oriented Architecture (SOA) is one of the latest architectural style whose goal is to achieve loose coupling among interacting software agents. SOA is a system architecture model consisting of standardized components such as web services[10]. A service is a unit of work done by a service provider to achieve desired end results for a service consumer. Both provider and consumer are roles played by software agents on behalf of their owners. In an SOA, resources are made available to other participants in the network as independent services that are accessed in a standardized way[11,12].

SOA and cloud computing are related, specifically, SOA is an architectural pattern that guides business solutions to create, organize and reuse its computing components, while cloud computing is a set of enabling technology that services a bigger, more flexible platform for enterprise to build their SOA solutions. In other words, SOA and cloud computing will coexist, complement, and support each other. Together SOA and cloud can provide a complete services-based solution[13]. Service-Oriented Cloud Computing Architecture (SOCCA)is proposed so that clouds can interoperate with each other. SOCCA supports easy application migration from one cloud to another and service redeployment to different clouds by separating the roles of service logic provider and service hosting/cloud providers[14,15]. It promotes an open platform on which open standards, ontology are embraced. More specifically, SOCCA can be developed and deployed in NCSEs to enable electromechanical system to perform closed-loop diagnostics across levels of maintenance.

2.2. Integrated Diagnostics Architecture

In order to overcome problem of existing equipment support information systems technology architecture diversification, tight coupling between the parts and poor information sharing and reuse, an open system, net-centric architecture is desired. Basing on Service Oriented Cloud Computing Architecture, an open, loose-coupled architecture of O-Level integrated diagnostics is put forward. It is four-layer structure as shown in Figure 1, including the physical resource layer, service resources layer, service bus layer and service application layer.

(1) Physical resources layer: this is the basic layer which provides hardware support for the architecture, including physical resources such as PMA, remote cloud diagnostics platform and communication network systems.

(2) Service resources layer: this is the layer that includes many fine-grained services. The services can be packaged as a single service or integrated into a coarse-grained service. According to different levels and different demands of services, the services can be divided into four categories: Data as a Service, Software as a Service, Platform as a Service and Infrastructure as a Service.

① Data as a Service (DaaS): Data such as product data, use information, fault knowledge, test data can be provided on demand to maintainer of all levels as a service. It can effectively support troubleshoot and optimize design of equipment by increasing effective accumulation, sharing, reuse of information of the full life cycle.

② Software as a Service (SaaS): Software such as test program, fault diagnosis software, IETM, data analysis software can be provided on demand to O-Level maintainer as a service. The software are shared by multiple users, automatically updated from the clouds. The maintainers can log in cloud diagnostics platform to download and install according to requirement. It saves the maintainers from the troubles of software deployment and maintenance.

③ Platform as a Service (PaaS): It provides a development platform with a set of services to assist application design, development, testing, deployment, monitoring, hosting on the cloud. It can support geographically distributed O-Level maintainer to work on projects collaboratively.
④ Infrastructure as a Service (IaaS): It virtualizes computing power, storage and network connectivity of the data centers, and offers it as provisioned services to O-Level maintainer. The maintainer can scale up and down these computing resources on demand dynamically. Typically, multiple users coexist on the same infrastructure resources.
The dividing lines for the four services are not distinctive. Components and features of one service can also be considered to be in another service. For example, data storage service can be considered to be either in IaaS or PaaS.
(3) Service bus layer: this is system connecting hub, is an essential element to build the nervous system of a maintenance organization. It is mainly to complete the functions of service management, information transformation, information routing, protocol conversion involved in service publication, request and offer.
(4) Service application layer: this is provide maintainers of all levels, technical support center, spare parts warehouse, production units and other client with specific service, such as test data analysis, fault diagnosis, remote expert diagnosis and decision-making services.

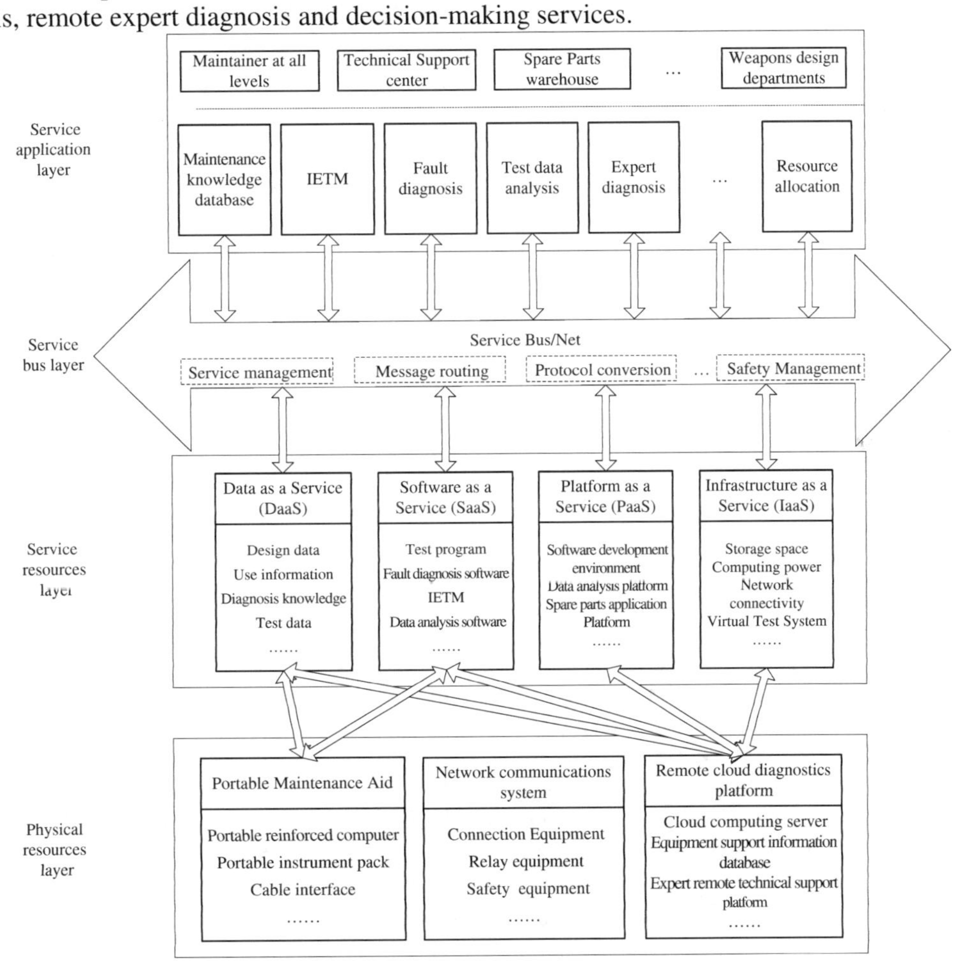

Figure 1. O-Level Integrated Diagnostics Architecture

3. O-Level Integrated Diagnostics System Model

According to O-Level integrated diagnostics architecture, the technical scheme of O-Level IDS for electromechanical system was put forward basing on network-centric architecture and standard diagnostics and maintenance information description. Test and diagnosis resources of repair shop at all levels and total life cycle information resources of electromechanical system were integrated by PMA with cloud diagnostics platform. The model of O-Level IDS for electromechanical system was developed as illustrated in Figure 2.

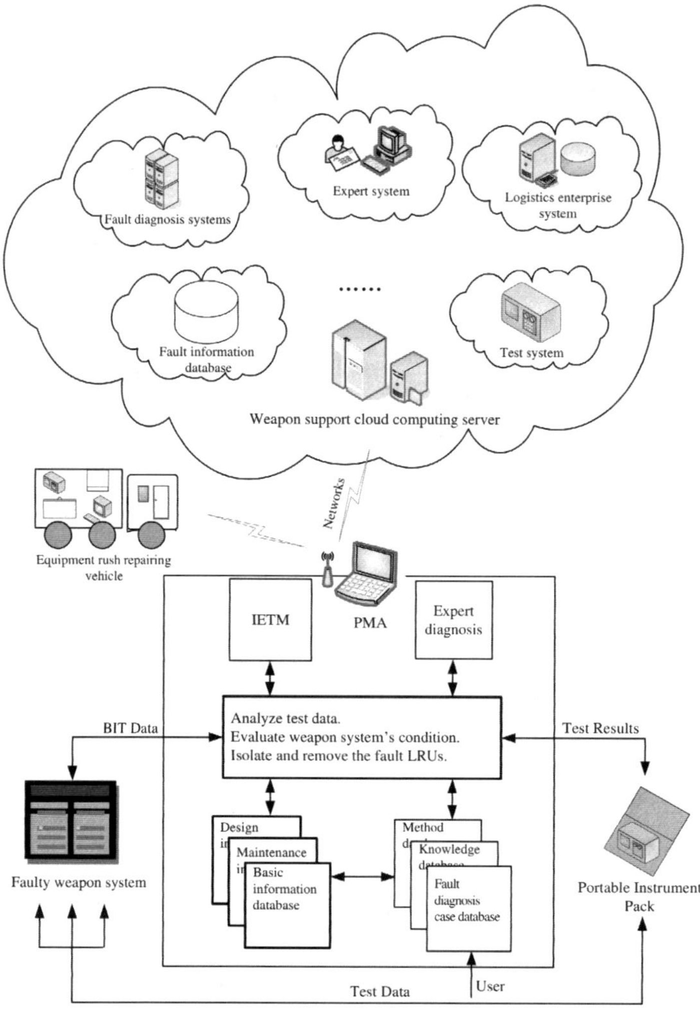

Figure 2. The model of O-Level Integrated Diagnostics System

When an electromechanical system is subject to failure, the first approach for fault diagnostics and trouble shooting is the analysis of self-monitoring data that acquire by Built In Test Equipment (BITE). The fault diagnostics can be support by Interactive Electronic Technical Manual (IETM), maintenance information database, fault diagnosis case database. If maintenance personnel can't locate and isolate failure, further functional signal acquisition and data analysis can be carried out by Portable Instrument Pack. If necessary, request support remote technical support as quickly and accurately detect the presence of electromechanical system hidden faults, fault isolation replaceable unit. Direct, real-time, and engineering assistance can be enabled when needed.

O-Level IDS that seeks to coordinate distance support with real-time test data, permitting local domain experts to assist forward deployed technicians. Hence, ready availability of information referenced resources, such as a collaborative historical maintenance data, usage data, the latest test methodologies and repair techniques, along with the latest experience can provide necessary leverage to execute a timely repair and return a problem electromechanical system to service more quickly.

4. Conclusion

IOP Conf. Series: Materials Science and Engineering **626** (2019) 012001 doi:10.1088/1757-899X/626/1/012001

In this paper, we put forward an open, loose-coupled architecture of O-Level IDS for electromechanical system basing on Service-Oriented Cloud Computing Architecture (SOCCA). The model of O-Level IDS was developed. Test and diagnosis resources that are integrated by PMA with cloud diagnostics platform are divided into four categories: data as a service, software as a service, platform as a service and infrastructure as a service. This allows more complex maintenance issues to be accomplished at FOLs by better utilizing the most relevant diagnostic information to enable direct, real-time, and engineering assistance when needed. O-Level IDS is capable of assisting O-Level maintainer to make accurate and timely decision, so it can effectively improve O-level maintainer support ability of complex electromechanical system.

References

[1] Cao, Jianfu, J. Zhang, and W. Tian. Fault diagnosis of large-scale electromechanical system based on nonlinear spectrum. *2016 35th Chinese Control Conference (CCC) IEEE*, 2016.

[2] IEEE, Bwmeta. Element. Diagnostic of AC machine based complex electromechanical systems. *Conference of the IEEE Industrial Electronics Society* 2011.

[3] Mi, Jinhua. Reliability assessment of complex electromechanical systems under epistemic uncertainty. *Reliability Engineering & System Safety*, vol. 152, pp 1-15, 2016.

[4] Zhou Y, Hui X. Fault diagnosis method of large-scale complex electromechanical system based on extension neural network. *Cluster Computing*, vol. 2, pp 1-10, 2018.

[5] Stora M, Kalgren P, *Remote Intelligent Diagnostics for Electronic Systems (RIDES), IEEE AUTOTESTCON*, 2009, pp 11-16,2009.

[6] Qian W U, Liu D F, Study and application of the integrated diagnosis technology for the equipment, *China Shiprepair*, vol. 6, pp 50-52, 2008.

[7] Xiangjun S, Long L, Junrong J, IETM technology research in the Integrated Diagnostics of the Equipment System, presented at *the Computer Application and System Modeling (ICCASM), 2010 International Conference on IEEE*, pp 643-645, 2010.

[8] Department of Defense, *Open Systems Approach Integrated Diagnostics Demonstration (OSAIDD) Study Final Report*-1999.

[9] Liu Jingyong, Integrated Diagnostics Applied in the Weapon System Supports in the US Army, Measurement & Control Technology, vol. 11, pp 1-3, 2004.

[10] Freschi S, Cost and benefit considerations for implementing an open systems approach to integrated diagnostics, presented at the AUTOTESTCON'99. *IEEE Systems Readiness Technology Conference*, pp 391-404, 1999.

[11] Zimmer D, Rhodes D, Human-machine interfaces, Industry Applications Magazine IEEE, vol. 1, pp 29-35, 2006.

[12] Embedded system design: Embedded systems foundations of cyber-physical systems, Marwedel P, *Springer Science & Business Media*, 2010.

[13] Papazoglou M P, Service-oriented computing: Concepts, characteristics and directions, presented at the Web Information Systems Engineering, 2003, *Proceedings of the Fourth International Conference on. IEEE*, pp 3-12, 2003.

[14] Alwardt A L, Nielson A R, Utilizing a service-oriented architecture to perform closed-loop diagnostics in network centric support environments, presented at the *IEEE Autotestcon*, pp 332-339, 2007.

[15] Tsai, Wei Tek, S. Xin, and J. Balasooriya. Service-Oriented Cloud Computing Architecture. *Seventh International Conference on Information Technology: New Generations* 2010.

IOP Conf. Series: Materials Science and Engineering **626** (2019) 012002 doi:10.1088/1757-899X/626/1/012002

Research on Arc Voltage Distance-Regulating System for TIG Welding Robot

Xuan Sun[a*],Zhiyong Wang[b], Yingjun Li and Changsheng Ai

School of Mechanical Engineering, University of Jinan, Jinan, China

[a] me_sunx@ujn.edu.cn, [b] 971917165@qq.com

Abstract. In order to meet the demand of TIG welding robot for high precision and high real time, this paper presents the arc voltage distance-regulating system using expert controller. By precisely controlling the distance between the tungsten electrode and the welding seam, the arc voltage is precisely controlled, and the closed-loop control of arc voltage is realized. The simulation and experiment of MATLAB show that the arc voltage distance-regulating system's arc voltage control accuracy is 18±0.8V(6±0.32mm), which meets the precision requirements of welding robot's arc voltage control and significantly improves the welding quality.

1. Introduction

TIG welding is widely used in thin plate welding. In order to ensure the quality of welding, the arc voltage must be stable within a certain range. In the process of TIG welding robot welding thin plate, the thin plate will produce large thermal deformation, welding leakage, welding nodule and other welding defects, so it must use the arc voltage distance-regulating system to adjust the welding arc voltage. Since the deformation of the thin plate during welding is nonlinear, the arc voltage variation is also nonlinear. Therefore, the arc voltage distance-regulating system of TIG welding robot cannot adopt the common PID control algorithm[1], so this paper adopts the expert PID control algorithm[2].

The relationship between arc voltage and arc length during TIG welding can be expressed as $U_h=U_z+K\cdot L$. In the equation, Uh is the arc voltage; U_z(U_z=3V) is the sum of cathode voltage drop and anode voltage drop under a certain condition (a certain current and electrode material); K(K=0.4) represents the proportional constant, arc voltage gradient; L is the distance from the tungsten to the weld. It can be seen that the arc voltage is dependent on the change in the length of the arc. When the arc is elongated, the arc voltage rises, and as the arc length becomes shorter, the arc voltage decreases.

2. Overall design of the arc voltage distance-regulating system

During the process of sheet welding, the distance between the welding gun and the surface of the sheet changes constantly due to the thermal deformation of the sheet, which results in the change of arc voltage (arc voltage is directly proportional to the distance between the welding gun and the welding surface). Because the stable arc voltage has a very important influence on the welding quality, the stability of arc voltage in the welding process can greatly improve the welding quality. Therefore, this paper designs the arc voltage distance-regulating system. The arc voltage distance control system is mainly composed of LMS filter and expert controller. The expert controller is divided into expert PID controller and speed controller of distance-regulation. The real-time performance of the arc voltage distance-regulating system can be effectively improved by the speed controller in the expert controller[3]. The welding arc voltage can be stabilized in a reasonable range through the reasonable coordination of these two parts.

IOP Conf. Series: Materials Science and Engineering **626** (2019) 012002 doi:10.1088/1757-899X/626/1/012002

First, the arc voltage acquisition board of the arc voltage distance-regulating system (as shown in Fig.1) compresses the collected arc by 10 times. Then, the arc voltage is subjected to LMS filtering to obtain a smoothly varying arc voltage. Finally, the smooth arc voltage is transmitted as an input signal to the expert controller to control the stepping motor, and the distance from the tungsten electrode to the weld is adjusted to achieve arc voltage regulation.

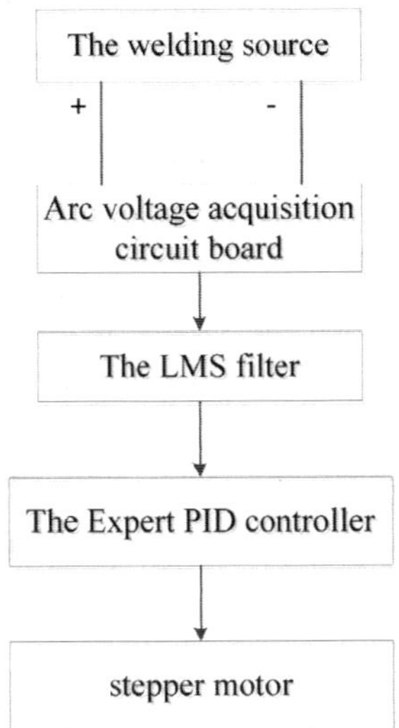

Figure 1. The arc voltage distance-regulating system

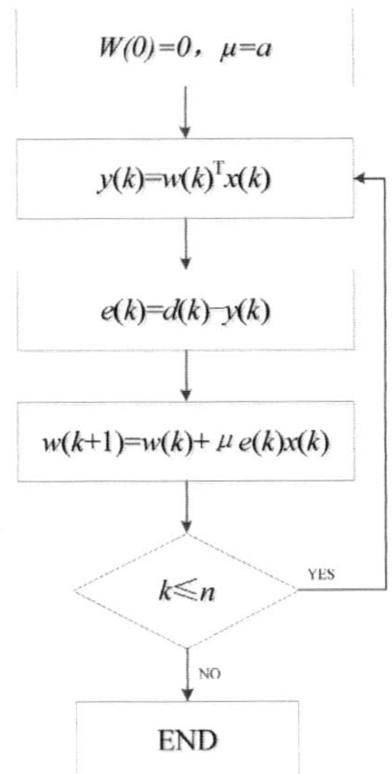

Figure 2. The LMS filtering algorithm

LMS filter is actually a special wiener filter that can automatically adjust its own parameters[4]. It does not need to know the statistical characteristics of the input signal and noise in advance when calculating, but can gradually estimate the required statistical characteristics in the process of work, and automatically adjust its own parameters based on this to achieve the best filtering effect[5]. Once the statistical characteristics of the input signal change, it can track the change, automatically adjust the parameters, so that the filter performance to achieve the best[6].

The LMS filtering algorithm is shown in Fig.2. In the equation, $x(k)$ is the input signal; $d(k)$ is the reference signal; $e(k)$ is the error signal; μ is the convergence factor; $w(k)$ is the weight; y(k) is the output signal. Output smooth signal after n iterations. Since the calculation process of the convergence factor is more complicated, in the actual application process, it is generally tested according to the actual situation, and the convergence factor with better filtering effect is obtained. The convergence factor is obtained through the MATLABsimulated analysis.

After filtering, the arc voltage signal will be transmitted to the expert controller, and the expert intelligent control is established on the knowledge of the controlled object and control law, without knowing the accurate mathematical model of the controlled object, and using the expert experience to design the parameters in the system. Expert intelligent control is a kind of direct expert controller. In the direct expert control system, the control signal is given directly by the system, which affects the whole controlled process. The direct expert control system derives the control signal at each sampling time according to the process information and rules in the knowledge base. As shown in Fig.3, the expert intelligent control system is composed of input and output sampling signals, knowledge base, information acquisition and processing, reasoning mechanism and control rule set. The expert controller processes the sampled signals, and then repeatedly matches them with the control rules in the knowledge base to obtain the corresponding results, and then controls the controlled objects for output.

Table 1. The expert PID control rules

$e(k)$	$\Delta e(k)$	$\Delta e(k-1)$	$u(k)$
$[E_{max}, +\infty]$			U_{max}
(E_{set1}, E_{max})	≥ 0		$u(k-1)+k_1[k_p \cdot \Delta e(k)+k_i \cdot \Delta e(k)+k_d \cdot \Delta^2 e(k)]$
	< 0	> 0	$u(k-1)+k_1 k_p e_m(k)$
		≤ 0	$u(k-1)$
$(0, E_{set1})$	≥ 0		$u(k-1)+k_p \cdot \Delta e(k)+k_i \cdot \Delta e(k)+k_d \cdot \Delta^2 e(k)$
	< 0	> 0	$u(k-1)+k_2 k_p e_m(k)$
		≤ 0	$u(k-1)$
0			$u(k-1)$
$(E_{set2}, 0)$	> 0	≥ 0	$u(k-1)$
		< 0	$u(k-1)+k_2 k_p e_m(k)$
	≤ 0		$u(k-1)+k_p \cdot \Delta e(k)+k_i \cdot \Delta e(k)+k_d \cdot \Delta^2 e(k)$
(E_{min}, E_{set2})	> 0	≥ 0	$u(k-1)$
		< 0	$u(k-1)+k_1 k_p e_m(k)$
	≤ 0		$u(k-1)+k_1[k_p \cdot \Delta e(k)+k_i \cdot \Delta e(k)+k_d \cdot \Delta^2 e(k)]$
$(-\infty, E_{min})$			U_{min}

E_{max}, E_{min}, E_{set1}, E_{set2} are the deviation boundary, and $E_{min} < E_{set2} < 0 < E_{set1} < E_{max}$; k_1 is the gain factor, and $k_1 > 1$; k_2 is the inhibiting factor, and$0 < k_2 < 1$; k_p, k_i, and k_d are the proportional, integral, and differential coefficients of the PID control, respectively.

Table 2. The speed control rule of distance-regulation

$e(k)$	$V_Y(k)$	$V(k)$
$[E_{max}, +\infty)$	$[V_{set2}, V_{max}]$	V_1
	(V_{set1}, V_{set2})	V_2
	$(0, V_{set1}]$	V_3
(E_{set1}, E_{max})	$[V_{set2}, V_{max}]$	V_4
	(V_{set1}, V_{set2})	V_5
	$(0, V_{set1}]$	V_6
$(0, E_{set1}]$	$[V_{set2}, V_{max}]$	V_7
	(V_{set1}, V_{set2})	V_8
	$(0, V_{set1}]$	V_9
0	$[V_{set2}, V_{max}]$	V_{10}
	(V_{set1}, V_{set2})	V_{11}
	$(0, V_{set1}]$	V_{12}
$(E_{set3}, 0)$	$[V_{set2}, V_{max}]$	V_7
	(V_{set1}, V_{set2})	V_8
	$(0, V_{set1}]$	V_9
$(E_{set3}, E_{set2}]$	$[V_{set2}, V_{max}]$	V_4
	(V_{set1}, V_{set2})	V_5
	$(0, V_{set1}]$	V_6
$(E_{min}, E_{set3}]$	$[V_{set2}, V_{max}]$	V_1
	(V_{set1}, V_{set2})	V_2
	$(0, V_{set1}]$	V_3

E_{max}, E_{min}, E_{set1}, E_{set2} are the deviation boundary, and $E_{min} < E_{set3} < E_{set2} < 0 < E_{set1} < E_{max}$; V_{max}、V_{set1}、V_{set2} is the speed boundary, and $0 < V_{set1} < V_{set2} < V_{max}$.

3. Establishment of the arc voltage distance-regulating system algorithm

Let $e(k)=U_{set}-U(k)$, In the equation, U_{set} is the optimal welding arc voltage of TIG welding robot; $U(k)$ is the actual welding arc voltage , which is collected by the arc voltage acquisition board at the k-th sampling time; $e(k)$ is the deviation between the actual welding arc voltage and the optimal welding arc voltage, which is measured at the k-th sampling time; $e(k\text{-}1)$ and $e(k\text{-}2)$ are the deviations between the actual welding arc voltage and the optimal welding arc voltage, which is measured at the $(k\text{-}1)$-th and $(k\text{-}2)$-th sampling moments, in addition

$$\left.\begin{aligned} \Delta e(k) &= e(k)-e(k-1), \\ \Delta e(k-1) &= e(k-1)-e(k-2), \\ \Delta^2 e(k) &= e(k)-2e(k-1)+e(k-2). \end{aligned}\right\}$$

In the equation, $\Delta^2 e(k)$ is the second-difference of the arc voltage deviation, which is measured at the k-th sampling time; $\Delta e(k)$ and $\Delta^2 e(k)$ represent the trends of $e(k)$ and $\Delta e(k)$, respectively; $V_Y(k)$ is the welding speed; $V(k)$ is the stepping motor speed of the arc voltage distance-regulating system. $e(k)$ and $\Delta u(k)$ are the input and output of the k-th sampling moment of the expert PID controller respectively. This paper proposes the expert PID control rules of the arc voltage distance-regulating system, as shown in Table 1. $e(k)$ and $V_Y(k)$ are the inputs of the speed controller of distance-regulation, and $V(k)$ is the output. This paper proposes the speed control rule of distance-regulation for the arc voltage distance-regulating system, as shown in Table 2.

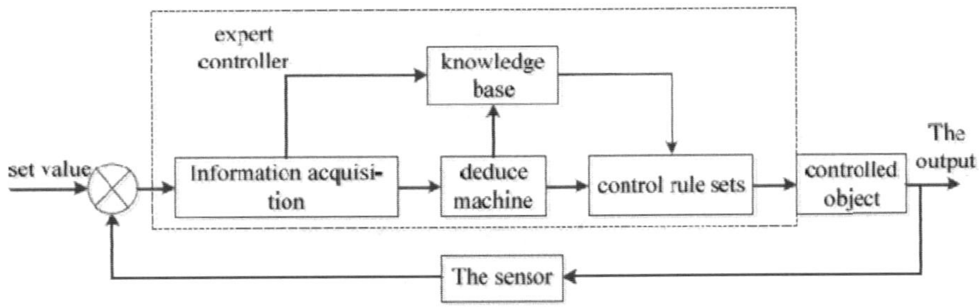

Figure 3. The arc voltage distance-regulating system

4. Simulation experiment

In order to test the performance of LMS filtering algorithm and expert PID control algorithm in the arc voltage distance-regulating system, this paper uses MATLAB to simulate and analyze these two algorithms, and the simulation results are shown in Fig.4 and Fig.5. In Fig.4, "input" is the 1000 arc voltage values collected by the arc voltage acquisition board; "output" is the output result obtained by the LMS filtering algorithm. It can be seen from Fig.4 that the LMS filtering algorithm can better remove the interference signal and fit the curve, which can effectively ensure that the signal amplitude is not weakened. In Fig.5, "out" is the output arc voltage value after adjusting by the arc voltage distance-regulating system; "set" is the optimal welding arc voltage; "actual" is the unregulated arc voltage value. It can be seen from Fig.5 that the arc voltage can be stabilized within the allowable error range through the adjustment of the expert PID controller and the speed controller of distance-regulation.

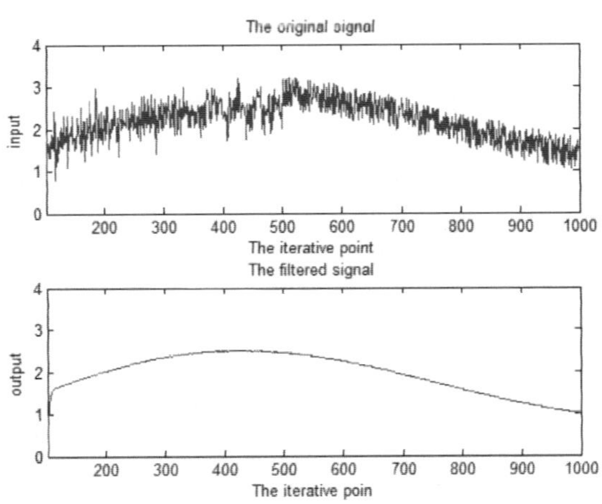

Figure 4. The LMS filtering algorithm

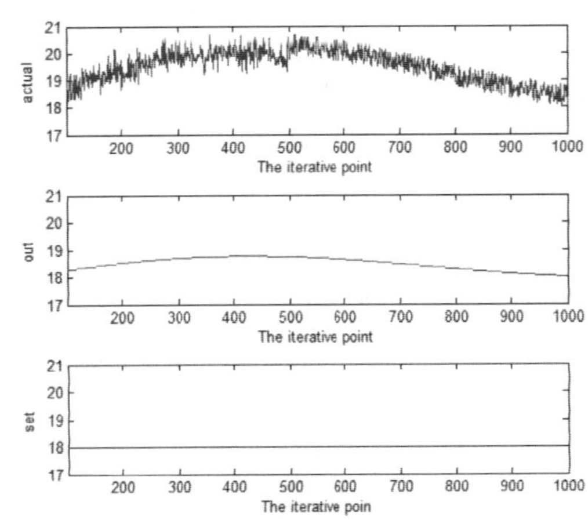

Figure 5. The Expert controller

In order to further test the performance of the arc voltage distance-regulating system, this paper uses the arc voltage acquisition board to collect 1000 regulated arc voltage values and normal arc voltage values. The statistical results are shown in Fig. 6. According to the experimental results, the value of

IOP Conf. Series: Materials Science and Engineering **626** (2019) 012002 doi:10.1088/1757-899X/626/1/012002

welding arc voltage adjusted by the arc voltage distance-regulating system is closer to the optimal value of arc voltage than the normal value of arc voltage, and is stable within the allowable error range(18±0.8V).Therefore, the distance between the tungsten electrode and the weld bead is stable at (6±0.32)mm, meeting the working requirements of TIG welding robot.

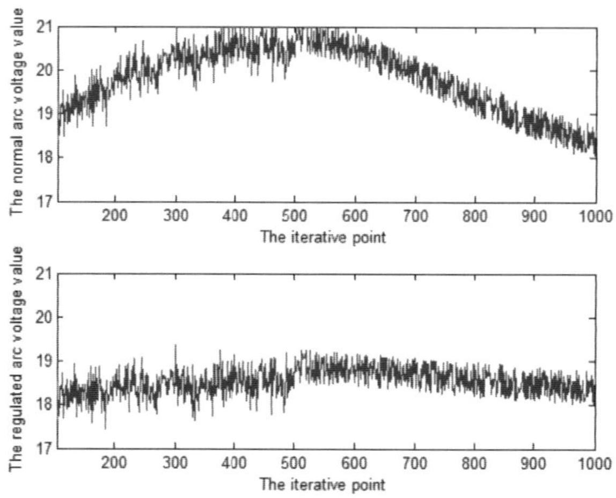

Figure 6. The data ofcollected arc voltage

5. Conclusion

Through the adjustment of the arc voltage distance-regulating system, the arc voltage is kept within the accuracy range of ±5%, which satisfies the welding quality requirements of the welding robot.

Acknowledgment

This work is supported by the national natural science foundation project (the project number is 51875250)

References

[1] Shi Dequan, GaoGuili, GaoZhiwei, Xiao Peng, Application of Expert Fuzzy PID Method for Temperature Control of Heating Furnace[J]. Procedia Engineering, 2012.

[2] QianZhengZai, Xin Gong Cai, Tao Jin Niu, Predictive Control Based on Fuzzy Expert PID Tuning Control[J]. Advanced Materials Research, 2012 (466).

[3] JianXiong Long, Research on Modeling and Simulation of Expert PID Controlled Servo System Based on MATLAB/S-Function[J]. J Applied Mechanics and Materials, 2013 (347).

[4] Zhi Sheng Jing, Ze Long Zhou, Chen Mei, Xiang Yong Su, Zhuo Yang, Yi Tao, Weight Size Determined Variable Step Size LMS Method for Identifying under Damped Systems[J]. Applied Mechanics and Materials, 2014 (511).

[5] Jun Sun, CaiHui Song, Xiao Hua Xiao, Xia Ming Jin, JiHeng Ni, A Variable Step Size LMS Adaptive Filtering Algorithm and Simulations Based on Computational Verb Theory[J]. Advanced Materials Research, 2013 (756).

[6] Tian Fu Qing, LuoRong, A Novel Variable Step Size LMS Algorithm Based on Modified Hyperbolic Tangent and Its Simulation[J]. Advanced Materials Research, 2012 (490).

IOP Conf. Series: Materials Science and Engineering **626** (2019) 012003 doi:10.1088/1757-899X/626/1/012003

Design of Answer-Grabbing Devices Based on Labview

Yang Zhao[1], Wu Rui*[2], Wang Youjia[1], Li Yuan[1]
[1]Electronic communication engineering college, Anhui Xinhua University, Hefei, 230088, China;
[2]Civil and environmental engineering college, Anhui Xinhua University, Hefei, 230088, China

47936560@qq.com

Abstract. Answer-grabbing devices are widely used in various TV programs, competitions and other fields. At present, with the increasing demand for fairness in competition, higher requirements have been put forward for answer-grabbing devices, such as higher sensitivity, better reliability, more powerful functions and so on. To meet this requirement, this paper designs a answer-grabbing devices based on Labview, which consists of answer-grabbing module, display module, compere button module and control module. Through design of front panel and program, debugging and operation, the functions of answer-grabbing device, as timing, scoring and zero clearing are finally realized. Experiments show that the system has high reliability, sensitivity and multiple functions.

1. Introduction

Nowadays, with the diversification of information, various contests begin to appear in the shows. As knowledge contests, Lyric guessing contests, etc. In these contests, competitors are often required to compete for the right to answer questions. The general method is that after the compere reads the questions, he will send out the signal to start answering, and then the competitor who responds to the signal fastest will get the right to answer questions, if only by relying on the compere's on-site judgment, it is easy to cause misjudgment, so human beings invented the answer-grabbing devices. The answer-grabbing devices should be stable, accurate and interactive in order to improve the overall level and ornamentation of the competition.

At present, all the answer-grabbing devicess on the market are equipped with physical equipment, which is difficult to install and debug, expensive and unstable. In order to solve the above problems, this paper designed and developed a new type of answer-grabbing devices with the technology of virtual instrument.

2. System design

The system consists of answer-grabbing button, start button, control center, display for score, display for time,timer, pilot lamp and setting device. We can set the time of answering questions by setting device. The system starts to run after the compere presses the start button. When a competitor presses the answer first button, the corresponding pilot lamp turns green. At this time, the competitor enters the Q & A session. And the other competitors can't answer any more, then the timer starts to run and display time. The first competitor can answer the question before the end of the time. The compere judges whether the competitor's answer correctly or incorrectly, press the True button or False button

to add or subtract points. If the competitor grabs the right to answer questions, but he does not answer within the time limit, the corresponding pilot lamp turns red.
The system chart is shown in Figure 1.

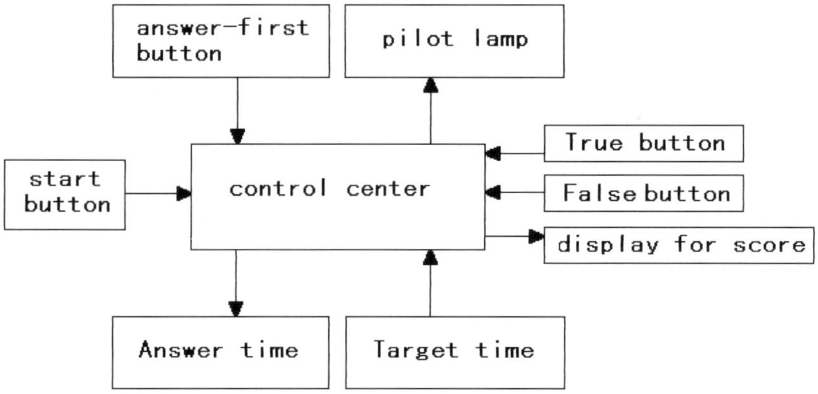

Figure 1. System Chart

This scheme has many functions, such as answering, timing, scoring, zero clearing, and so on. Because it does not involve latches and decoders, the system is relatively simple and has good stability.

3. Front panel design of the system

The front panel design of the system is divided into five modules, namely the answer-grabbing button module, the display for score module, the compere button module, the pilot lamp module and the answer time module.
According to the number of participants, the number of answer-grabbing buttons is different. In this program,the answer-grabbing buttonsr are selected from the front panel-control-Boolean-confirmation button. This design adopts the scheme of four keys, which can help four competitors to race at the same time, and reserve space for expanding into multi-channels answer-grabbing devices.
Labview provides a variety of display modules, it is simple to call and intuitive to display. Considering the intuitive requirement of the score display in this design, the numerical direct display scheme is adopted. The numeric display is selected from the front panel - control - numeric - numeric display control, and each competitor corresponds to a numeric display.
By consulting the information, the compere buttons should include start button, correct button, error button, zero-setting button, and each button is selected from the front panel-control-Boolean-confirmation button.Among them, the start button is used to start the whole system. After the start of the system, the following functions such as timing, answering and scoring can be performed. And only when the start button is pressed, the competitor's answer-grabbing button can be pressed, otherwise, the answering is invalid. In other words, when answering in advance, there is no response when the button is pressed. After the competitor race to be the first to answer a question successfully and answer questions, the compere judges the correctness and error of the answers. If the answer is correct, the compere presses the correct button, and the competitor gains points; if the answer is incorrect, the compere presses the incorrect button, and the competitor loses points (or no points).At the end of a round of game, the zero-setting button is used to clear the scores, timing and initialize the current status of the answer-grabbing buttons and the pilot lamps in the round. The exit button is used to finish and exit the program.
The pilot lamps have two display states: red light and green light, (red light means answering time-out, green light means racing to be the first to answer a question successfully), each pilot lamp is selected from the front panel-control-Boolean-circular indicator light.The system consists of four pilot lamps.
Answer time module includes answer time limit device and answer time display. Display control is selected from front panel - control - numeric - numeric input control.

IOP Conf. Series: Materials Science and Engineering **626** (2019) 012003 doi:10.1088/1757-899X/626/1/012003

Answer time limit device is used to set the time limit for a competitor to answer questions. If the answer is timeout, the corresponding red light in front of the competitor lights up.
Answer time display is used to display the time spent by competitor. It is realized by front panel-control-numerical-numerical display control.
The front panel of the answer-grabbing devices is designed as shown in Figure 2.

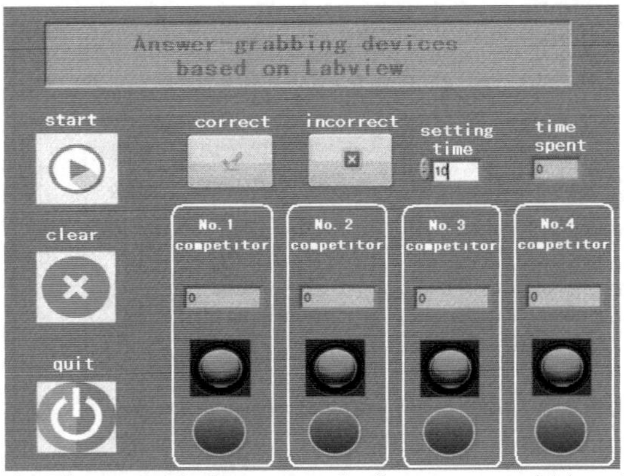

Figure 2. Figure with front panel of answer-grabbing devices

4. Program panel design of the system

This design first initializes the program, then enters the key detection, and when it detects that the start button is pressed, it will start to answer. Detection status of the four competitors' answer-grabbing buttons and the compere's zero-setting button. If a competitor press the answer-grabbing button, he will start to answer question. If the zero-setting button is pressed, the time and score will be cleared, this round of game will finish.
The system flowchart is shown in Figure 3, and the program of the system is shown in Figure 4.

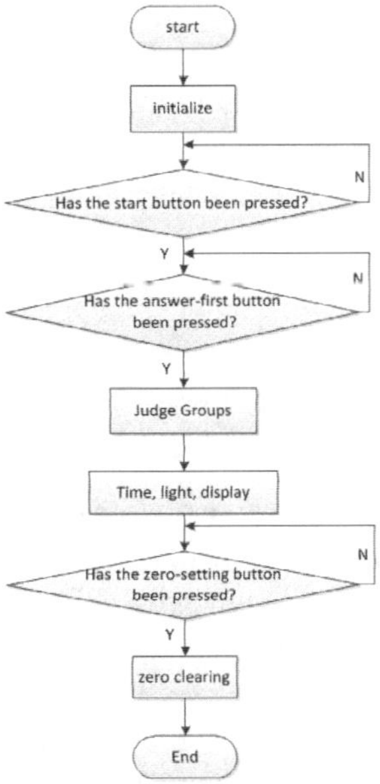

Figure 3. System flowchart

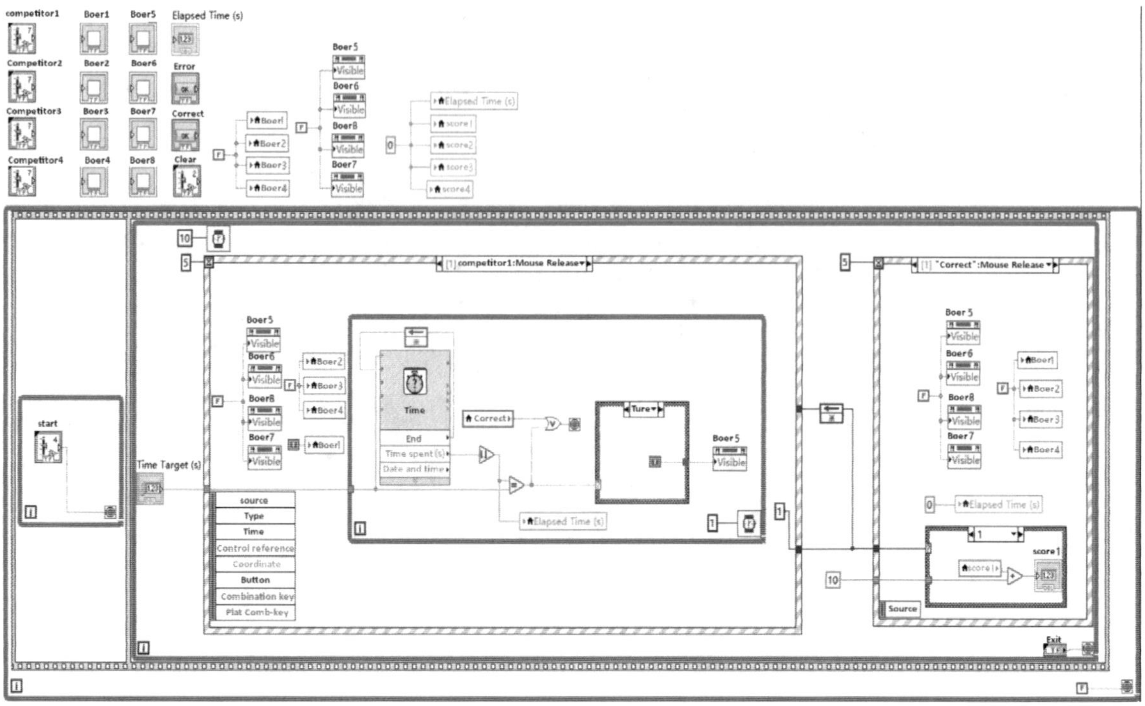

Figure 4. Program of the system

5. Debugging and Operation

After the front panel and program design of the system are finished, it can enter the debugging stage. The specific debugging operation is shown below.

5.1. Operating state I

Under normal working conditions, a competitor raced to be the first to answer a question successful after the compere pressed the start button, and answered correctly within a limited time, the compere should press the correct button to add 10 points to the competitor, as shown in Figure 5.

5.2. Operating state II

The competitor raced to be the first to answer a question successful after the compere pressed the start button, but answered incorrectly, the compere should press the incorrect button, the corresponding pilot lamp turns red, as shown in Figure 6.

5.3. Operating state III

After a round of fierce response, the compere is responsible for recording the data, resetting and initializing the system, as shown in Figure 7.

Figure 5. Normal working state of the system

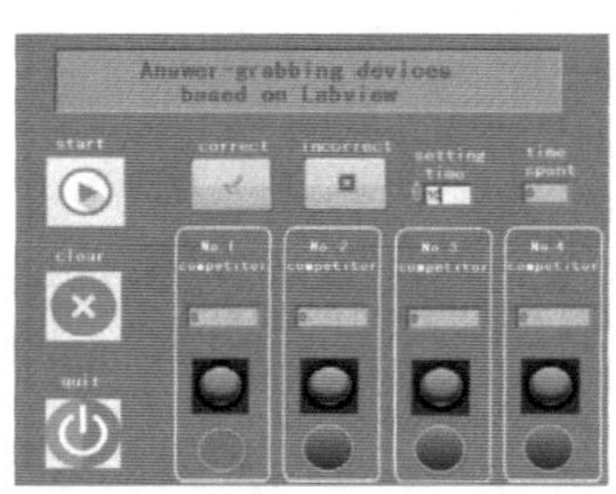

Figure 6. Timeout state of the system

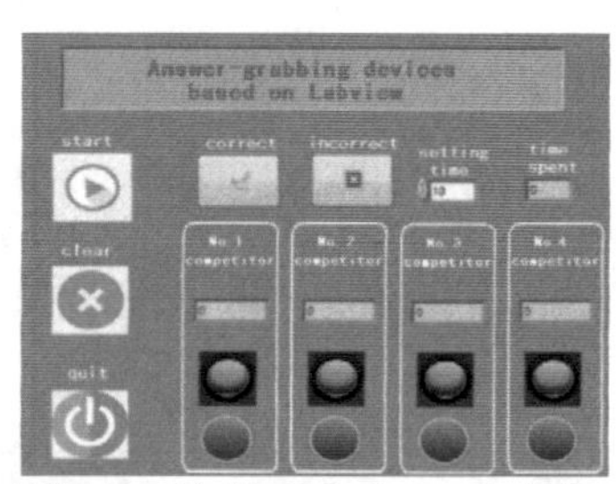

Figure 7. Reset state e of the system

IOP Conf. Series: Materials Science and Engineering **626** (2019) 012003 doi:10.1088/1757-899X/626/1/012003

6. Conclusion

Aiming at the demand of current market, we designs a answer-grabbing devices based-on Labview, which has the functions of answering, timing display, scoring display and compere control on-site, it can meet the requirements in most environments.

In the later stage, we will continue to work hard to add the hardware to be a part of the system, it will make the system more perfect.

Acknowledgments

This work was supported by Anhui province colleges and universities "quality engineering" project No.2018sysxx02, the Anhui Natural Science Research Project No.KJ2017A616, and the domestic outstanding young backbone talents in universities of visiting program No. gxgnfx2018070. National Students Innovation Project No. 201812216032

7. References

[1] Butler Frank K and Giebner Stephen 2019 *Journal of special operations medicine : a peer reviewed journal for SOF medical professionals* 9(2)

[2] Jyoti Khatri KC,Gerard Fitzgerald and Meen B. Poudyal Chhetri 2019 *Prehospital and Disaster Medicine* 34(3)

[3] Li Hongqiang,Meng Jiangbin,Qu Baoju,Dong Xiaojuan,Lou Guangjun and Huang Binqi 2019 *Journal of Shandong University of Technology(Natural Science Edition)* 5-9

[4] Butler Frank K,Giebner Stephen. Tactical 2019 *Journal of special operations medicine : a peer reviewed journal for SOF medical professionals* 19(2).

[5] Fu Guohong,Zhang Qianfeng and Cheng Hui 2019 *Progress in Geophysicsni* 1-10

[6] Zhang Liangshan, Ma Shuai and Qin Tao 2017 *Light Engineering Technology* 86-87

[7] Wang Kun 2018 *Electronic Design Engineering* 94-97

[8] Wang Yalei and Cao Jing 2017 *Chinese Patent*

[9] Liu Xin and Jin Xuehong 2017 *Software Guide* 135-137

[10] Wang Qiang and Jiao Ji 2017 *Taxation* 194

IOP Conf. Series: Materials Science and Engineering **626** (2019) 012004 doi:10.1088/1757-899X/626/1/012004

A Hybrid Piezoelectric and Inductive Rotational Energy Harvester

Jacob Bianconi, Jesse Hallett, Jessica Pealo and Reza Rashidi

Department of Mechanical and Electrical Engineering Technology, State University of New York, Alfred State College, Alfred, NY 14802, USA

rashidr@alfredstate.edu

Abstract. This paper presents the development of a rotational energy harvester that generates a power based on a combination of piezoelectric and electromagnetic methods. The device utilizes two piezoelectric beams and two inductors and potentially multiple elements on a stationary plate to maximize the harvested energy. The piezoelectric beams generate voltages when deflected with a rotating magnet. In addition, the inductors generate voltages when a magnet passes over them. In order to maximize the harvested energy and frequency, sixteen magnets were embedded in the rotating plate to excite the elements with a higher frequency. A 3D printer was used to make the enclosure, and stationary and rotating components of the device. An assembly of Arduino microcontroller, DC motor, IR sensor and LCD screen was used to test the device. The harvester utilized a closed loop feedback system to monitor rotational speed. The device was tested at various rotational speeds from 130 rpm to 510 rpm. Voltage output versus time graphs were analyzed for each speed. It was determined that raising the number of moving magnets and rotational speed increase not only the frequency but also the voltage amplitude in both piezoelectric beams and inductors. The developed energy harvesting device can be used in any existing rotational system such as vehicle tires, bicycles, rotating machines and exercise equipment.

1. Introduction

Energy harvesting has been widely researched in the past decade due to its great usage for providing energy to remote areas as well as electronic devices [1]. Harvesting energy from piezoelectric beams is a popular form of energy conversion due to the possibility for a high voltage output [2]. Piezoelectric energy harvesting devices have been used for charging batteries [3], wireless remote power supply [4], subsurface ocean/river power generators [5], unmanned aerial vehicles [6], biomedical devices [7], and other portable devices [8]. There have been numerous studies that have focused on producing electrical energy from natural vibration in sources such as the human body and other wearable devices [9].

One of the available energy sources that are already existent and natural is rotational movement. This kinetic energy can be potentially converted into electrical energy to power remote devices and not rely on batteries [10]. This practice of extracting power for every day systems that rotate has been applied to rotational systems such as bicycles. The converted energy from the wheels on the bicycle could be used to power accessories the user has on the bicycle which would remove the need for a battery pack [11]. Many rotational energy harvesting operate at low frequencies such as frequencies between 7-13.5 Hz [12]. They compared two different configurations of magnets at higher frequencies and the power outputs. In addition, different shaped cantilever beams including rectangular and trapezoidal were compared to find the maximum deflection and potentially highest power output [13].

Piezoelectric based rotational energy harvesters have been demonstrated on systems such as wind systems [14], automotive tires [15], wearable human watches [16], rotary disks [17], rotational sun track solar panels [18] and wireless autonomous monitoring systems [19]. Theoretical and experimental modelling of a piezoelectric based rotational energy harvester with bi-stability and frequency up conversion was demonstrated for applications with wide bandwidth and low-frequency kinetic energy [20]. Furthermore, models have been characterized for energy harvesters in self-powered watches [21,22]. Few rotational energy harvesters have been demonstrated to use multiple PZT beams. For example, a wrist-worn rotational energy harvester was developed that used six PZT beams with four moving magnets [23]. The eccentric rotor-based piezoelectric rotational energy harvester was developed based on the idea adopted from self-winding watches to address low performance challenges in energy harvesting from human motion. The design worked based on both linear and rotational motions and provided several degrees of freedom [24]. In another research, several feasible magnetic plucking configurations were categorized and simulated for eccentric energy harvesters. The work classified in-plane plucking and three different out-of-plane deflection configurations including direct repulsive, orthogonal, and indirect repulsive setups [25]. Rotational energy harvesters utilizing multiple piezoelectric beams and moving magnets were demonstrated [26,27].

An electromagnetic vibration-driven energy harvester was developed using printed circuit board materials to lower the cost and to achieve lightweight device [28]. An electromagnetic energy harvesting technique with an enhanced magnetic flux change was presented based on Faraday's law to convert mechanical vibration into electrical energy using an array of alternating north- and south-orientation magnets [29]. In this paper, we have added multiple inductors to the piezoelectric based rotational energy harvesters to increase the power output with a hybrid piezoelectric and inductive method. The paper descriptively discusses the design, fabrication, testing and results of the rotational energy harvester actuated by magnetic fields to harvest energy from a rotational movement. In addition, the design uses different elements such as a closed loop feedback system to control the rotational speed of the system. The effect of rotational speed on the harvester performance including voltage amplitude and frequency are discussed.

2. Design

Energy conversion into electrical power using inductive and piezoelectric methods require some type of kinetic energy. The energy in this design is a rotational movement. A spinning plate passes magnets over the two different types of energy generation elements, inductors and piezoelectric cantilever beams, converting a rotational energy into an electrical power. A complete view of the system is depicted in figure 1.

There are two plates in the device, rotational and stationary. The top plate, the one rotating, has two rings of magnets at different radii. The inner ring with a 23 mm radius has eight 4.8 mm diameter magnets which pass over the piezoelectric beams. In addition, the outer ring with a 47.5 mm radius has eight 12.7 mm diameter magnets for the inductors. The number of magnets increase the operational frequency that acts on the elements. Instead of one pass per revolution there are eight passes per revolution simulating a faster rotational speed by eightfold. The polarity of all the magnets are in the same direction north to south and perpendicular to the plate surface. Figure 2 shows the top and bottom plates in detail, including the two rings of magnets in the top plate, and the PZT beams and the inductors in the bottom plate. The bottom plate houses the energy conversion elements, two piezoelectric beams and two inductors. The 34.85 mm long, 2.50 mm wide, 0.30 mm thick cantilever beams are made of lead zirconate titanate (PZT). The two PZT beams which are 90° apart have a neodymium magnet on one end, therefore when the top plate magnets pass by, the repulsion between two magnets causes a deflection in the beam. The distance between the two magnets is a crucial part of the system. If the distance is too large, the beams do not bend enough to generate a voltage while if it is too small the repulsion force largely bends the beam and breaks it.

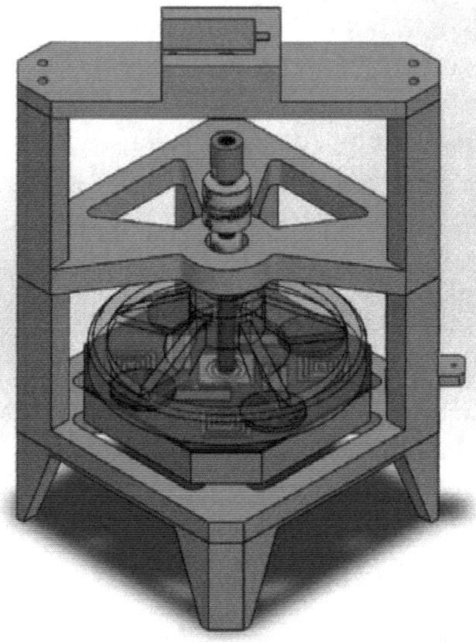

Figure 1. Complete view of the inductive and piezoelectric rotational energy harvester.

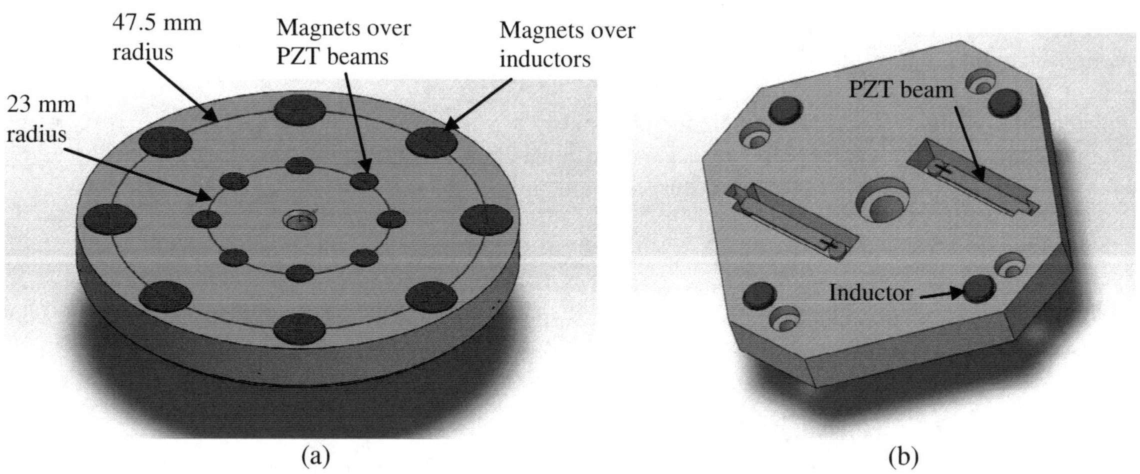

Figure 2. (a) The top rotating plate including two rings of magnets, eight 12.7 mm diameter magnets centered over inductors and eight 4.8 mm diameter magnets centered over PZT beams. (b) The bottom stationary plate including the PZT beams and the inductors. A small magnet is attached to the end of each PZT beam.

Working principle of the piezoelectric and inductive energy harvesting is illustrated in figure 3. The piezoelectric actuation is based on repulsive force between two magnets on the beam and rotating plate. As the rotating magnets rotate along with top plate, they line up with the magnet on the PZT beam at some point and deflect the beam downwards. As the rotating magnet gets away from the beam magnet, the beam deflects back to the original position. The beam will continue deflecting upwards in negative direction due to spring elastic forces. The signal generated by the PZT beams is a sinusoidal

IOP Conf. Series: Materials Science and Engineering **626** (2019) 012004 doi:10.1088/1757-899X/626/1/012004

pulse and the voltage output is related to the beam deflection. When the beam reaches its max point of deflection, the voltage is at its greatest.

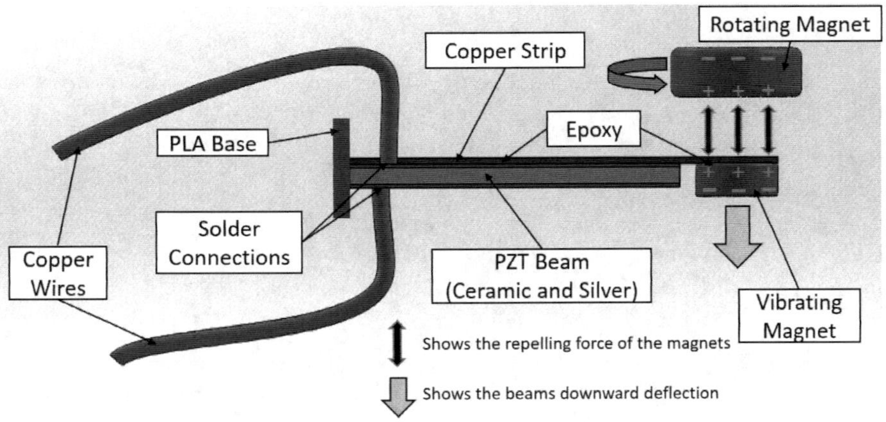

(a)

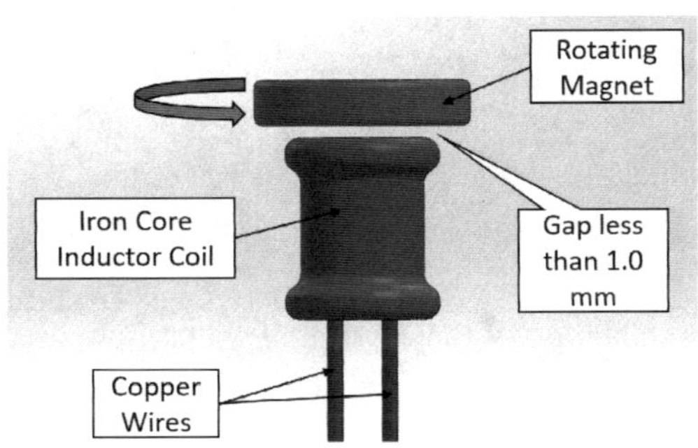

(b)

Figure 3. Working principal of piezoelectric beam (a) and inductive (b) energy harvesting.

The inductive energy harvesting principal is based on Faraday's law which states that an electromotive force is induced in a conductive coil by time varying magnetic flux though the coil. The change in magnetic field in this device is made by moving the magnet on the rotating plate over the inductor. The vertical gap between the rotating magnet and the 10mH inductors needs to be at the minimum possible value to receive the highest voltage. This gap is designed to be about 1 mm. The inductors are designed to be 7.2 mm diameter to ensure that the 12.7 mm moving magnets completely cover the inductors when passing over them.

3. Fabrication

The design was realized using SOLIDWORKS®. The device frame, holding the rotating magnets, PZT beams, inductors, DC motor and IR sensor was fabricated using an Athorbot® A-01 3D printer using poly-lactic acid (PLA) filament (Hatchbox®). The utilization of 3-D printing allowed for features and components to be modeled with little down time. A complete overview diagram of the fabricated device and its accessories is shown in figure 4.

IOP Conf. Series: Materials Science and Engineering **626** (2019) 012004 doi:10.1088/1757-899X/626/1/012004

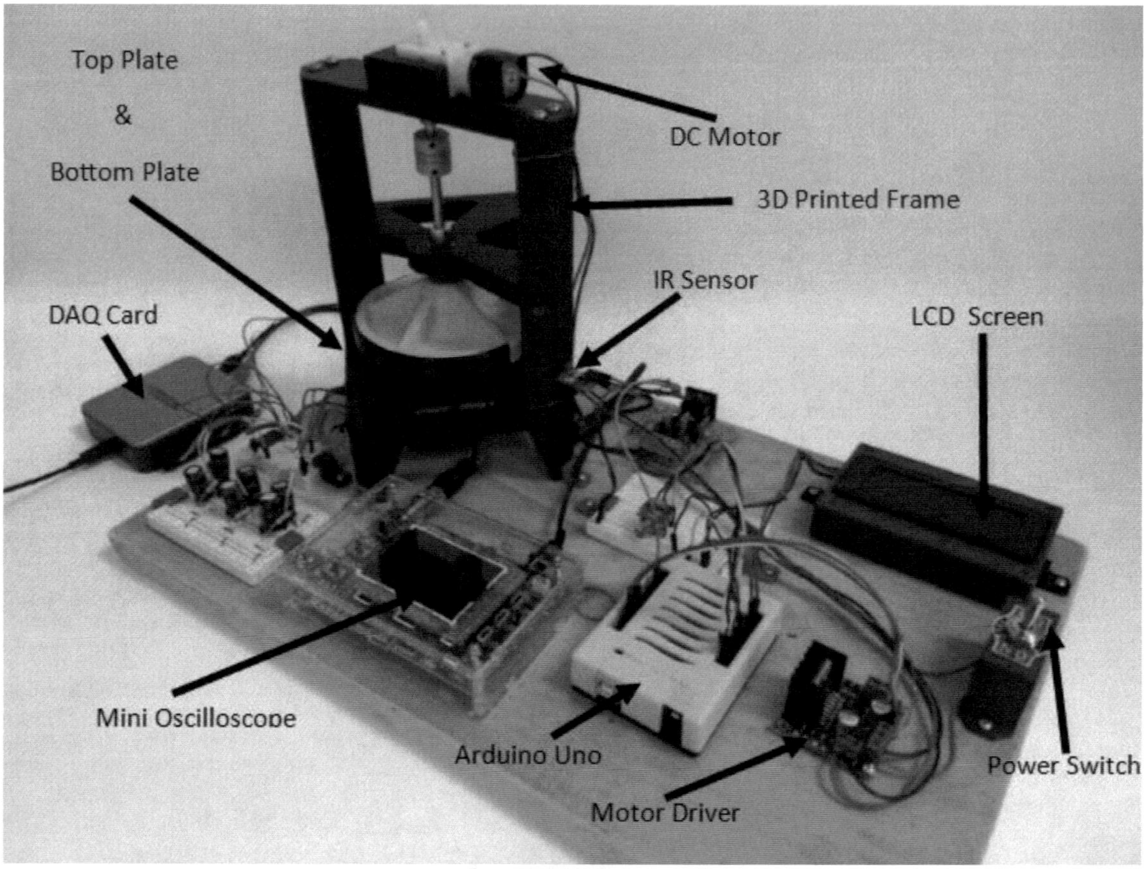

Figure 4. An overview diagram of the fabricated energy harvester and its accessories.

The rotating magnets were adhered to the underside of the rotating plate in which its vertical distance from the harvesting elements could be adjusted. This was done by adding or subtracting 0.1mm thick shims to move the shaft that the rotating wheel was fixed to up and down. The 6 mm diameter shaft was held in place by three ball bearings, two of which were pressed into the plate that mounted the inductors to ensure that the magnets rotated perfectly parallel and at a constant distance from the surface of the inductors. The driving motor was mounted above the shaft and a torque was transmitted via an aluminum flexible shaft collar to ensure that the motor did not bind due to any misalignment.

The two piezoelectric beams were laminated to a thin piece of narrow copper sheet on the upper side of the piezoelectric beam. There was a magnet mounted on one end of the beam via the copper sheet using an epoxy. The beams were bonded into a well inside the 3D-printed base plate with the PZT beam facing down to ensure that it was firmly fixed to the base. This forced the beam to deflect both the copper and the PZT material when a magnet passed over the end of it. Wires of the off-the-shelf inductors and PZT beams were soldered to the outside of the bottom plate through the designed holes. Figure 5 shows a PZT beam and an inductor fully fabricated and in place on the bottom plate.

Figure 5. Close-up image of a PZT beam and an inductor fully fabricated and in place on the bottom plate.

A closed-loop feedback control system was used to control the rotational speed of the top plate. An Arduino Uno and a DC motor driver board acted as the controller and the actuator, respectively. The DC motor connected to a shaft and spun the top plate. A pulse-with-modulation (PWM) output signal was used to control the DC motor driver. A set rotational speed was measured from a potentiometer and the Arduino converted the analog input signal to a set speed range. An infrared (IR) sensor was used to measure the actual rotational speed by sending a digital signal to the Arduino where a set of mathematics were done to convert the digital input into the speed. The set speed, the actual speed and the PWM signal were all displayed on an LCD utilizing I2C communication. A code was programed in Arduino to control the PWM based off the set speed and the measured speed. Arduino used library references to set up the known functions in a script.

4. Testing

The device was run using its aforementioned accessories and tested using a National Instruments USB 6003 Multifunctional Data Acquisition (DAQ) device to make a precise measurement of each energy harvesting element. A LabVIEW software was used to allow for the recording of generated voltages as a function of time. To fully test the system, data was collected at six distinct speeds from 130 rpm to 510 rpm. This range of speeds are available in many applications such as fitness equipment, washing machine, bicycle, motorcycle and car tires.

5. Results and Discussion

In order to analyze changes in the output of each harvesting element, various speed measurements including those at 130 rpm, 250 rpm, 300 rpm, 400 rpm, 435 rpm and 510 rpm were taken with the DAQ card. Figures 6 and 7 show typical generated voltages as a function of time at two rotational speeds, 130 rpm and 510 rpm for PZT beams and inductors, respectively. It is clear that due to the use of eight magnets on the moving plate for both PZT beams and inductors, the frequency of the sinusoidal voltages significantly increased compared with a case if only one moving magnet was used. The cycles were repeated faster by the number of magnets, making the multiple magnet case a significantly higher overall output. In addition, the rounds of cycles were repeated faster with higher rotational speed in which next round of cycles began earlier by a certain moving magnet. The latter effect was enhanced by having eight rotating magnets in a row. Therefore, the higher rotational speed, the higher overall output.

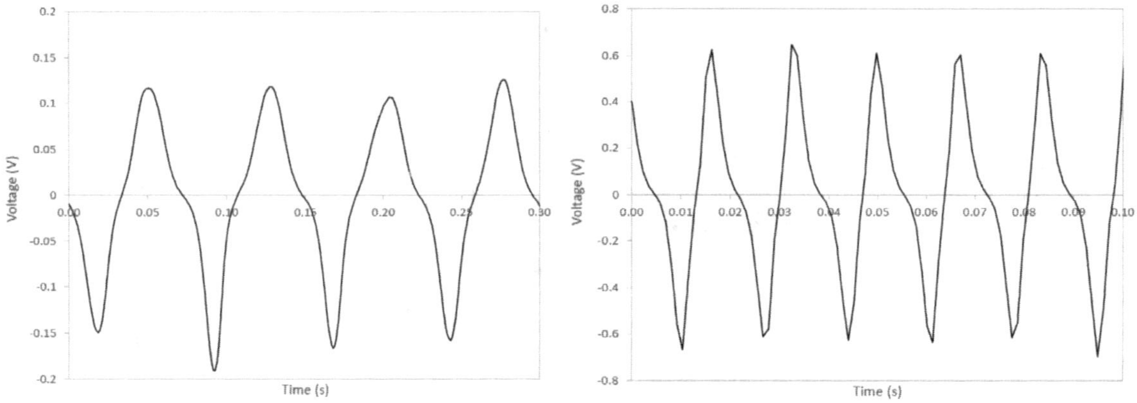

(a) (b)

Figure 6. Typical voltages generated by the PZT beams as a function of time at 130 rpm (a) and 510 rpm (b).

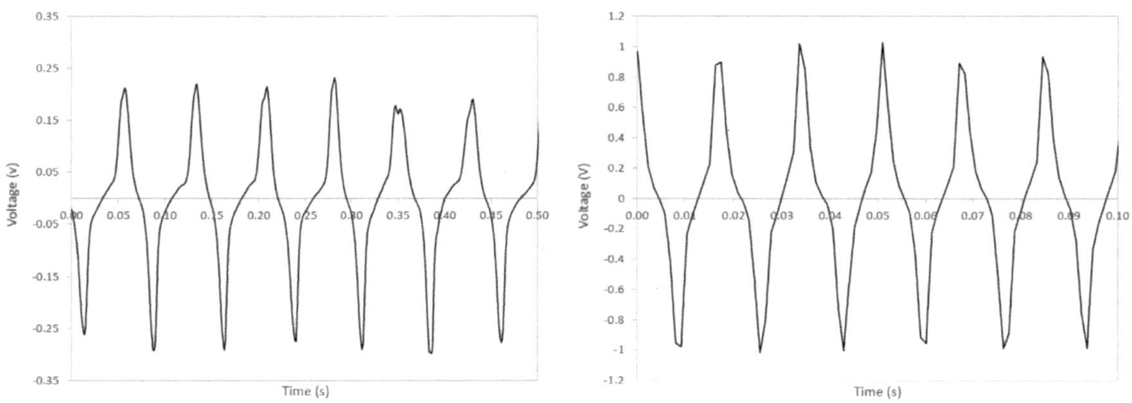

(a) (b)

Figure 7. Typical voltages generated by the inductors as a function of time at 130 rpm (a) and 510 rpm (b).

Figure 8 compares the average voltage amplitudes as a function of rotational speed for the PZT beams and inductors. The results showed that the average amplitude in both cases increased with the rotational speed. The voltages generated by the inductors were determined to be linearly proportional to the rotational speed while those generated by the PZT beams were found to be increasing with the rotational speed with a polynomial trend. Therefore, it is concluded that the efficiency of inductive energy conversion is higher than that of piezoelectric energy conversion at higher speeds. In addition, the inductors have more potential to generate voltages with higher amplitude than PZT beams in this case.

Furthermore, the harvester in this research was run with two piezoelectric beams and two inductors that increased the overall output. The harvester has a potential to be run with several piezoelectric beams and inductors that increase the overall output to a higher level. This is especially beneficial for applications that require higher voltages or draw a higher currents. In this case, the several piezoelectric beam circuits can be put in series or parallel, respectively in order to meet these requirements. A similar approach can be used for the inductors. Voltage cycles generated by the PZT beams or the inductors will have the same phases because the number of the moving magnets passing over the beams is even. In other words, when a beam is hit by a certain moving magnet, another beam 180° apart from the other beam is hit by a magnet 180° apart from the other magnet. This fact would be applicable for the inductors.

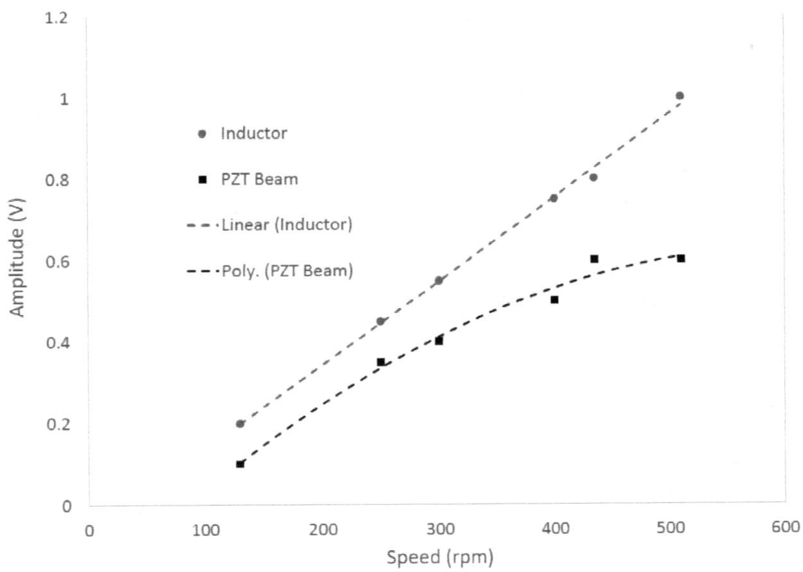

Figure 8. A comparison between the graphs of average voltage amplitude as a function of rotational speed for the PZT beams and inductors.

In addition, an AC/DC converter can be utilized to convert harvested voltages in each element to a DC voltage [30,31]. High voltages can be potentially used as a substitute for the grid power where a device requires that level of voltage. The required voltages can be adjusted by including different number of PZT beams or inductors. There are lots of such devices that draw a low current which is compatible with the voltage generated using this energy harvester. As an example, several monitoring devices in electronic screen of a stationary exercise bicycle are utilized to display speed, distance, time, calories burned, etc. An energy harvester can be implemented into the apparatus, and the power output of the user could be used to power the display. Though such an equipment may use internal converters to adjust the required voltages, the generated voltage by the energy harvester can be used to power the whole equipment unit.

6. Conclusions

The development of a hybrid piezoelectric and inductive energy harvester was demonstrated in this paper. The two methods simultaneously converted a rotational energy into an electrical one. It was found that the energy harvester's output increased by using the two alternative methods simultaneously. Using multiple moving magnets passing over the PZT cantilever beams and the inductors significantly increased the overall generated voltage and frequency. In addition, increasing number of the PZT beams and inductors largely increased the power output. Furthermore, efficiency of inductive energy conversion was higher than that of piezoelectric energy conversion at higher speeds. The inductors have more potential to generate voltages with higher amplitude than PZT beams in this case. The device could be utilized in any rotational machines such as vehicle tires, bicycles, watermills, windmills and exercise equipment. A further study could be performed to increase the efficiency of the beams using lead-free piezoelectric materials with the addition of rare earth metal via a chemical doping process. Furthermore, multiple inductors in different radial rings on the stationary wheel could be utilized to increase the power output.

References

[1] Erturk A, Inman DJ. *Piezoelectric energy harvesting*. John Wiley & Sons; 2011 Apr 4.
[2] Liu JQ, Fang HB, Xu ZY, Mao XH, Shen XC, Chen D, Liao H, Cai BC. A MEMS-based piezoelectric power generator array for vibration energy harvesting. *Microelectronics Journal.* 2008 May 1;39(5):802-6.
[3] Sodano HA, Park G, Leo DJ, Inman DJ. Use of piezoelectric energy harvesting devices for charging batteries. *InSmart Structures and Materials 2003: Smart Sensor Technology and Measurement Systems* 2003 Jul 22 (Vol. 5050, pp. 101-109). International Society for Optics and Photonics.
[4] Ottman G, Bhatt A, Hofmann H, Lesieutre G. Adaptive piezoelectric energy harvesting circuit for wireless, remote power supply. *In19th AIAA Applied Aerodynamics Conference* 2002 (p. 1505).
[5] Taylor GW, Burns JR, Kammann SA, Powers WB, Welsh TR. The energy harvesting eel: a small subsurface ocean/river power generator. *IEEE Journal of Oceanic Engineering.* 2001 Oct;26(4):539-47.
[6] Anton SR, Inman DJ. Vibration energy harvesting for unmanned aerial vehicles. InActive and Passive Smart Structures and Integrated Systems 2008 2008 Apr 18 (Vol. 6928, p. 692824). *International Society for Optics and Photonics.*
[7] Shi Q, Wang T, Lee C. MEMS based broadband piezoelectric ultrasonic energy harvester (PUEH) for enabling self-powered implantable biomedical devices. *Scientific reports.* 2016 Apr 26;6:24946.
[8] Cook-Chennault KA, Thambi N, Sastry AM. Powering MEMS portable devices—a review of non-regenerative and regenerative power supply systems with special emphasis on piezoelectric energy harvesting systems. *Smart materials and structures.* 2008 Jun 9;17(4):043001..
[9] Khaligh A, Zeng P, Zheng C. Kinetic energy harvesting using piezoelectric and electromagnetic technologies — state of the art. *IEEE Transactions on Industrial Electronics.* 2010 Mar;57(3):850-60.
[10] Anton SR, Sodano HA. A review of power harvesting using piezoelectric materials (2003 – 2006). *Smart materials and Structures.* 2007 May 18;16(3):R1.
[11] Vasic D, Chen YY, Costa F. Self-powered piezoelectric energy harvester for bicycle. *Journal of Mechanical Science and Technology*. 2014 Jul 1;28(7):2501-10.
[12] Manla G, White NM, Tudor MJ. Numerical model of a non-contact piezoelectric energy harvester for rotating objects. *IEEE Sensors journal*. 2011 Nov 11;12(6):1785-93.
[13] Baker J, Roundy S, Wright P. Alternative geometries for increasing power density in vibration energy scavenging for wireless sensor networks. *In3rd international energy conversion engineering conference* 2005 Aug 15 (p. 5617).
[14] Zhang J, Fang Z, Shu C, Zhang J, Zhang Q, Li C. A rotational piezoelectric energy harvester for efficient wind energy harvesting. *Sensors and Actuators A: Physical*. 2017 Aug 1;262:123-9.
[15] Hu Y, Xu C, Zhang Y, Lin L, Snyder RL, Wang ZL. A Nanogenerator for Energy Harvesting from a Rotating Tire and its Application as a Self‐Powered Pressure/Speed Sensor. *Advanced Materials*. 2011 Sep 15;23(35):4068-71.
[16] Lockhart R, Janphuang P, Briand D, de Rooij NF. A wearable system of micromachined piezoelectric cantilevers coupled to a rotational oscillating mass for on-body energy harvesting. *InMicro Electro Mechanical Systems (MEMS), 2014 IEEE 27th International Conference on* 2014 Jan 26 (pp. 370-373). IEEE.
[17] Larkin M, Tadesse Y. HM-EH-RT: hybrid multimodal energy harvesting from rotational and translational motions. *International Journal of Smart and Nano Materials*. 2013 Dec 1;4(4):257-85.

[18] Sathiyamoorthy S, Bharathi N. Hybrid energy harvesting using piezoelectric materials, automatic rotational solar panel, vertical axis wind turbine. *Procedia engineering.* 2012 Jan 1;38:843-52.

[19] Febbo M, Machado SP, Gatti CD, Ramirez JM. An out-of-plane rotational energy harvesting system for low frequency environments. *Energy Conversion and Management.* 2017 Nov 15;152:166-75.

[20] Fu H, Yeatman EM. Rotational energy harvesting using bi-stability and frequency up-conversion for low-power sensing applications: Theoretical modelling and experimental validation. *Mechanical Systems and Signal Processing.* 2019 Jun 15;125:229-44.

[21] Wang J, Wang W, Jewell GW, Howe D. Design of a miniature permanent-magnet generator and energy storage system. *IEEE transactions on industrial electronics.* 2005 Oct;52(5):1383-90.

[22] Pillatsch P, Yeatman EM, Holmes AS. A piezoelectric frequency up-converting energy harvester with rotating proof mass for human body applications. *Sensors and Actuators A: Physical.* 2014 Feb 1; 206:178-85.

[23] Xue T, Yeo HG, Trolier-McKinstry S, Roundy S. A wrist-worn rotational energy harvester utilizing magnetically plucked {001} oriented bimorph PZT thin-film beams. In 2017 *19th International Conference on Solid-State Sensors, Actuators and Microsystems (TRANSDUCERS)* 2017 Jun 18 (pp. 375-378). IEEE.

[24] Xue T, Yeo HG, Trolier-McKinstry S, Roundy S. Wearable inertial energy harvester with sputtered bimorph lead zirconate titanate (PZT) thin-film beams. *Smart Materials and Structures.* 2018 Jul 25;27(8):085026.

[25] Xue T, Roundy S. On magnetic plucking configurations for frequency up-converting mechanical energy harvesters. *Sensors and Actuators A: Physical.* 2017 Jan 1;253:101-11.

[26] Rashidi R, Summerville N, Nasri M. Magnetically Actuated piezoelectric-based rotational energy harvester with enhanced output in wide range of rotating speeds, in *IEEE Transactions on Magnetics.* 2019; doi: 10.1109/TMAG.2019.2918989.

[27] Rashidi R, Summerville N, Nasri M. A dual-purpose piezoelectric multi-beam energy harvesting and frequency measurement device for rotational applications, In *IEEE-MIT Undergraduate Research Technology Conference* 2018 Oct, Boston, MA

[28] Bouendeu E, Greiner A, Smith PJ, Korvink JG. Design synthesis of electromagnetic vibration-driven energy generators using a variational formulation. *Journal of Microelectromechanical Systems.* 2011 Apr;20(2):466-75.

[29] Zhang Q, Kim ES. Vibration energy harvesting based on magnet and coil arrays for watt-level handheld power source. *Proceedings of the IEEE.* 2014 Nov;102(11):1747-61.

[30] Bolt R, Magno M, Burger T, Romani A, Benini L. Kinetic AC/DC Converter for Electromagnetic Energy Harvesting in Autonomous Wearable Devices. *IEEE Transactions on Circuits and Systems II: Express Briefs.* 2017 Dec;64(12):1422-6.

[31] Peters C, Spreemann D, Ortmanns M, Manoli Y. A CMOS integrated voltage and power efficient AC/DC converter for energy harvesting applications. *Journal of micromechanics and microengineering.* 2008 Sep 29;18(10):104005.

IOP Conf. Series: Materials Science and Engineering **626** (2019) 012005 doi:10.1088/1757-899X/626/1/012005

Optimization Design Method for Hydrostatic Guideway Based on PM Restrictor

Wanjiao Dong, Beizhi Li, Chongjun Wu, Qinzhi Zhou, Weicheng Guo

Mechanical Engineering School, Donghua University, Shanghai 201620, China

lbzhidhu@163.com

Abstract. Stiffness and flow rate are always used to evaluate the performance of the hydrostatic slides and well-designed parameters, such as flow rate ratio, initial flow rate, pump pressure and oil film thickness, allow the slides achieving nanoscale precision. In this study, the relationship between the design parameters and the stiffness and flow rate is reveal by an analytical model. Then a multi-objective optimization problem is solved using Non-dominated Sorting Genetic Algorithm II. The optimization objectives are maximize stiffness and minimize flow rate. At last, the optimized design parameters are calculated and corresponding maximum stiffness and minimum flow rate of the hydrostatic slides are acquired.

1. Introduction

Nowadays hydrostatic guideways are widely employed in ultra-precision machines as high stiffness, high load carrying capacity, and low frictional losses at low-to-moderate speeds, minimal wear, good accuracy and excellent damping. Many scholars studied the method of design and optimization of hydrostatic guideways to improve their performance [1]. For hydrostatic bearing, several objectives are required to be considered, such as minimum consumption of power [2], maximum stiffness [3], maximum load carrying capacity [4] and maximum dynamic stiffness [4]. The minimization of the consumption of power is the objective most commonly considered. In the meanwhile, the optimization design principle of hydrostatic bearings is maximum rigidity and minimum flow. The improper selection of the design parameters may result in low stiffness, poor geometrical accuracy, poor surface finish and poor stabilization. This may affect the quality of the machine tool. Therefore, selection of optimal combination of design parameters for hydrostatic guideway is a matter of concern for designers and researchers. In most of the cases, designers analysed these design parameters by using the finite element method (FEM) [5] and selected optimum combination of design parameters through experience and traditional design method, which is costly, time consuming and tedious.

Thus, these factors have steered the designers towards applying optimization algorithm techniques for parameter design. Researchers had applied various classic optimization algorithms. However, the optimization problems of parameters design of hydrostatic guideway may not get the results using traditional methods. Therefore, an intelligent method is required for solving constrained design problems effectively.

In this work, a novel optimal design model of closed hydrostatic slide, which is controlled by PM (Progressive Mengen) restrictors, is proposed. Maximum stiffness and minimum flow rate are considered as the optimal objective, while the pump pressure, initial flow rate, flow rate ratio and oil film thickness are the design parameters. This multi-objective optimization problem is a constrained, non-linear and parameter bounded problem, which can be carried out by evolutionary algorithms.

IOP Conf. Series: Materials Science and Engineering **626** (2019) 012005 doi:10.1088/1757-899X/626/1/012005

Finally, the optimized design parameters and corresponding maximum stiffness and minimum flow rate are obtained.

2. Optimization design method of high-precision hydrostatic guideways

Based on the force balance equation, the load of the slide, F_i, which is original slides design from structure parameters, can be calculated. Figure 1 represents stress analysis of a couple of aspectant bearing pads. The stresses of the pad 1 and pad 3 are pr_1 and pr_3, respectively.

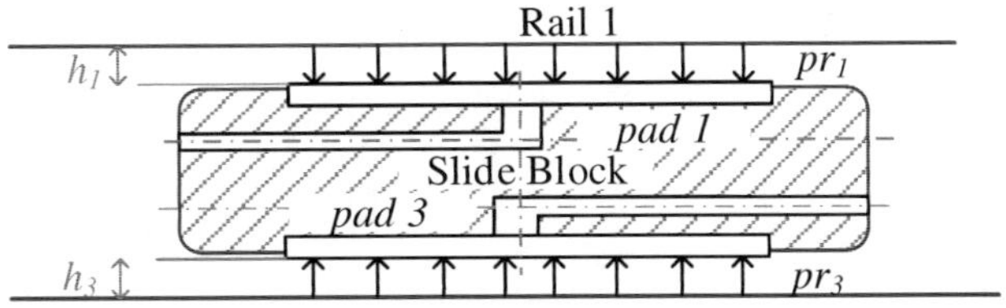

Figure 1. Stress analysis of a couple of bearing pads.

By mechanical analysis, the equation of a couple of aspectant bearing pads can be presents as the Equation 1. pr_i is the ith bearing pad pressure (i=1, 3). L_i and B_i are the length and width of the pad, respectively. l_i and b_i represents the length and width of the land, respectively. A_i is defined as effective area for pr_i and can be calculated by Equation 2.

$$F = pr_1A_1 - pr_3A_3 \tag{1}$$

$$A_i = (L_i - l_i)(B_i - b_i) \tag{2}$$

According to the calculation method of the flow rate in the pocket of the rectangular oil pad, the flow rate can be obtained as follow:

$$Q_{Pi} = \frac{pr_i}{R_i} \tag{3}$$

where R is defined as liquid resistance and can be calculated by $R_i = \dfrac{6\rho\gamma}{h_i^3\left(\dfrac{L_i - l_i}{b_i} + \dfrac{B_i - b_i}{l_i}\right)} = \dfrac{c_{Ri}}{h_i^3}$. ρ is oil density, γ is viscosity, and c_R is a constant depending on structure parameters of pads, oil density and viscosity.

For the hydrostatic slide which is using PM controller, the parameters of the controller to control the pressure of the bearing pads are as follows: flow rate ratio c_r, initial flow rate q_0, and pump pressure p_s. low rate ratio is calculated by $c_r = q_p / q_0$, where q_p represents the flow rate when the bearing pad pressure pr equals pump pressure p_s and q_0 represents the flow rate when the bearing pad pressure pr equals 0. The flow rate of bearing pads with PM controller can calculate using equation (4) in [7].

$$Q_{PMi} = q_{0i}\left[1 + (c_{ri} - 1)\, pr_i / p_s\right] \tag{4}$$

Assuming that there is no flow rate loss from the exit of PM restrictors to the pockets, the flow rate in the ith pocket should be equal to the flow rate out of the ith PM restrictor, which means $Q_{Pi}=Q_{PMi}$. Thus, the pad pressure and flow rate of the ith pocket can be computed by:

$$pr_i = \frac{p_s q_{0i} c_{Ri}}{p_s h_i^3 - (c_r - 1) q_{0i} c_{Ri}} \tag{5}$$

$$Q_i = \frac{p_s q_{0i} h_i^3}{p_s h_i^3 - (c_{ri} - 1) c_{Ri} q_{0i}} \tag{6}$$

Film stiffness, an important indicator to evaluate performances of hydrostatic slides, is the rigidity of the oil film, which shows the ability to resist changes from the external loads and deformation in response to an applied force. The stiffness is defined as:

$$ju = \sum_{i=I,II}^{i} ju_i = -\sum_{i=I,II}^{i} \frac{dF_i}{dh_i} = \sum_{i=I,II}^{i} \frac{3q_{0i}c_{Ri}A_i}{h_i^4\left(1-\frac{q_{0i}c_{Ri}\left(c_{ri}-1\right)}{h_i^3 p_s}\right)^2} \tag{7}$$

where the negative sign indicates that the changing trend of the load and the film thickness is the opposite.

3. Multi-parameter and multi-objective optimization algorithm

The principle of the optimization design of the hydrostatic bearing is well known as maximum stiffness and minimum flow rate. The optimization objectives are changed with the major parameters, such as, film thickness, pump pressure, initial flow rate and flow rate ratio.

Mathematical programming method always has low efficiency and high sensitive in weights or order of targets when the objective and constraint function of the multi-objective optimization problem are non-linear, non-differentiable, or discontinuity. At present, evolutionary algorithms are commonly applied to resolve multi-objective optimization problem [8].

In this paper, Non-Dominated Sorting Genetic Algorithm II (NSGA-II) [9] is used to solve the multi-objective optimization problems of the hydrostatic slides design. This algorithm maintains an elitist mechanism for maintaining good individual and improving the population evolution to keep population diversity, therefor, the algorithm is still widely used in the multi-objective optimization. The mathematical model are as follows:

$$\begin{cases} \min \quad F(x)=[f_1(x), f_2(x)] \\ \text{s.t.} \quad \begin{array}{l} h_{01} < x(1) < h_{02} \\ pr_1 < x(2) < pr_2 \\ c_{r1} < x(3) < c_{r2} \\ q_{01} < x(4) < q_{02} \end{array} \end{cases} \tag{8}$$

where $f_1(x) = -ju, f_2(x) = Q, x(1) = h_0, x(2) = pr, x(3) = c_r, x(4) = q_0$.

4. Case study

4.1. Boundary

According to the equation 4 and 5, the relationship between stiffness and flow rate and parameter p_S, q_0, h_0, and c_R are shown in Figure 2 to Figure 5.

In Figure 2, when the film thickness is less than 17μm, the flow rate increases rapidly then decreases. When the film thickness is 20μm, the change of flow rate remains stable. The flow rate changes rapidly that is not dependent on the motion stability of the slide.

In Figure 3, the stiffness reaches around 1400N/μm when pump pressure is 25bar and decreases slightly with pressure. The flow rate increases rapidly and reaches the maximum at when pump pressure is 18 bar. When the pressure keeps increasing, the flow rate decreases and later remains at 25 bar. In general, pump pressure is supposed to be higher than 25 bar, because the changes of stiffness and flow rate are unstable when the pressure is lower than 25 bar.

It can be overserved in Figure 4 that when the flow ratio is more than 3.5, stiffness and flow ratio changes dramatically. Thus, the flow ratio should be less than 3.5.

In Figure 5, when the initial flow is more than 0.015 l/min, the flow increases rapidly. The flow changed rapidly that is not dependable for the motion stability of the slide, therefore, the initial flow should be lower than 0.15 l/min.

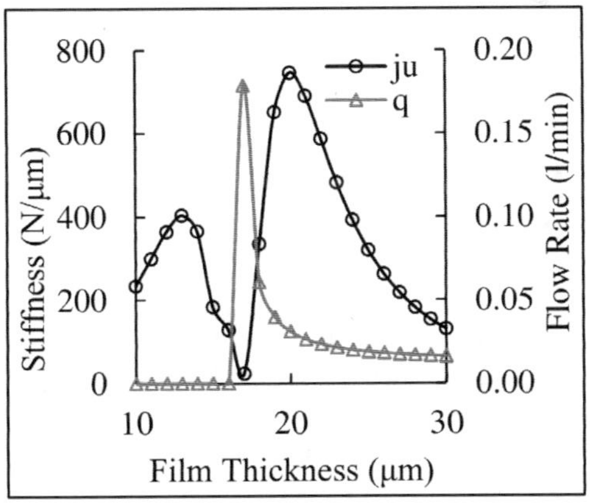

Figure 2. Relationship between film thickness and stiffness and flow rate.

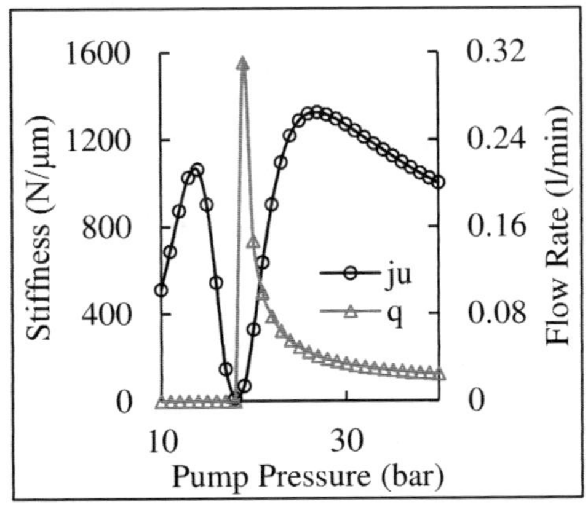

Figure 3. Relationship between pump pressure and stiffness and flow rate.

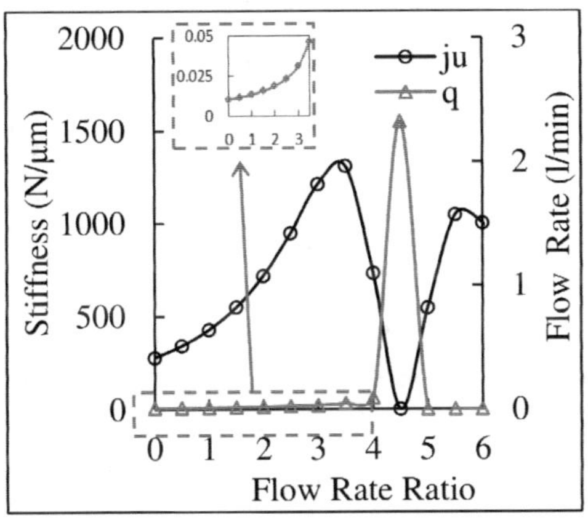

Figure 4. Relationship between flow rate ratio and stiffness and flow rate.

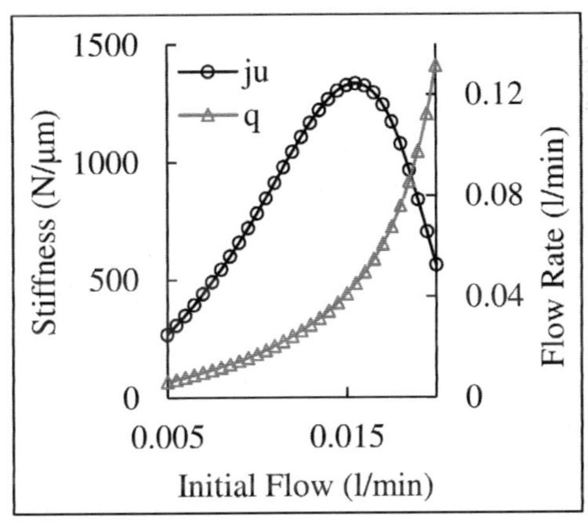

Figure 5. Relationship between flow rate ratio and stiffness and flow rate.

As a result, the range of the design parameters are:

$$\begin{cases} 18\mu m < h_0 < 30\mu m \\ 28bar < pr < 35bar \\ 2 < c_r < 5 \\ 0.01 l/min < q_0 < 0.015 l/min \end{cases} \tag{9}$$

4.2. Calculation and analysis

According to the optimal design model, the Pareto-front is shown in Figure 6. Ultimately, the selected design parameters of the hydrostatic slide are h_0 = 19 μm, pr = 3.2 MPa, cr = 2.5, and q_0 = 0.014 l/min. The optimal results are the stiffness ju=1237 N/μm and flow rate q=0.029 l/min. The stiffness ju is 1056N/μm and low rate q is 0.037l/min, respectively, while the original design parameters are oil film thickness h=20μm, pump pressure p_s=3.2MPa, initial low rate q_0=0.018 l/min., and flow rate ratio cr_5=2.75. Comparing the performance of the hydrostatic slides which were designed in different design parameters, it is observed that the performance of hydrostatic slides which was designed by the optimization design parameters is much better than the performance of hydrostatic slides which was designed by the original design parameters. The stiffness increases by 17% and the flow rate decreases by 22%.

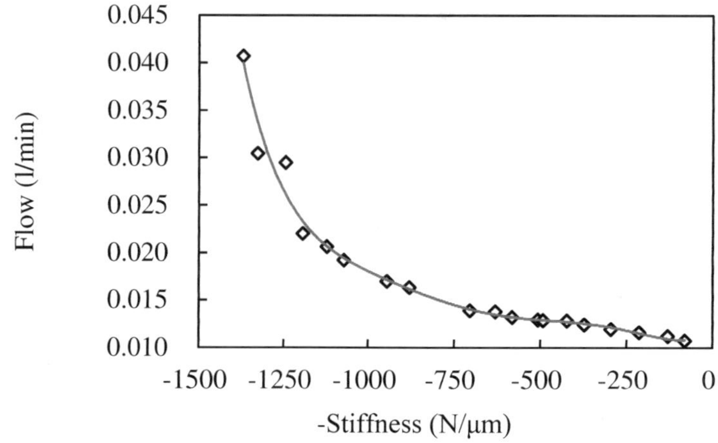

Figure 6. Pareto front of the optimized results.

5. Conclusions

In this study, a novel optimal design model is established for designing a high-performance hydrostatic slide. The effects of slide, oil supply system and restrictors on stiffness and flow rate are taken account into the proposed optimization model. To achieve the best performance of the hydrostatic slide, the objectives of the optimization problem are the maximum stiffness and the minimum flow rate, while the pump pressure, initial flow rate, flow rate ratio and oil film thickness are the design parameters. After optimization, the maximum stiffness is 1237 N/μm and minimum flow rate is 0.029 l/min when flow rate ratio, initial flow rate, pump pressure and oil film thickness are 2.5, 0.014 l/min, 32 bar and 19 μm, respectively. Compared with the hydrostatic slides optimally designed by traditional design method, the comprehensive performance of the hydrostatic slides designed by the proposed method shows overwhelming advantages. The stiffness increases by 17% and the flow rate decreases by 22%.

In the further research, the more effective and accurate evaluation method and intelligent optimization algorithm of hydrostatic slide should be proposed to provide better theoretical basis and design method for the design of ultra – precision hydrostatic slide.

References

[1] Fujita T, Matsubara A and Yamazaki K 2011 *Int J Mach Tool Manu* **51 (2)** 104-11.
[2] Wang Y, Zhao Y, Cai L, Liu Z, Cheng Q 2016 *J Vibroeng* **18 (7)** 4193-4206.
[3] Qiu M, Minson BR and Raeymaekers B 2013 *Tribol Int* **67** 278-88.
[4] Fesanghary M and Khonsari M. 2013 *Tribol Int* **67** 254-62.
[5] Shao J, Liu G, Yu X, Zhang Y, Meng X and Jiang H 2018 *Ind Lubr Tribol* **70 (1)** 68-75.
[6] He G, Sun G, Zhang H, Huang C and Zhang D 2017 *Int J Adv Manuf Tech* **93(5-8)** 1915-27.
[7] Yang S, Zhao P, Xu Y, Sun L, Wu P and Long X 2015 *Procedia CIRP* **27** 187-91.
[8] Jain H and Deb K 2014 *IEEE Transactions on Evolutionary Computation* **18 (4)** 602-622.
[9] Rong Y, Zhou Q, Huang Y, Chang Y, Zhang G and Shao X 2015 *Int J Adv Manuf Tech* **85 (5-8)** 1239-1247.

IOP Conf. Series: Materials Science and Engineering **626** (2019) 012006 doi:10.1088/1757-899X/626/1/012006

Development of Robotic Coordinated Control System for Bowl Food Packaging

Bao Yequan, Ai Changsheng*, Geng Dunyang, Ren Gangchang, Ju Hong

School of Mechanical Engineering, University of Jinan, Shandong Province, China

me_aics@ujn.edu.cn

Abstract. The processing of the bowled food is relatively mechanical, repetitive and cumbersome, resulting in high work intensity, low efficiency, and high defect rate, and meets the working cycle:13.5s, positioning accuracy: 2mm.A PLC-based automatic feeding control system for food packaging line was designed. Including the conveyor line for bowl feeding, the numerical controlled manipulator and package process following control system for reclaiming and discharging functions, the system control scheme was proposed and 3D modeling construction was carried out. The design of hardware and software for feeding, reclaiming and discharging in the system is elaborated in detail. Finally, through the actual on-site commissioning and production operation, it is proved that the equipment has fully met the design requirements and greatly improved the production efficiency.

1. Introduction

In modern society, with the prosperous development of the economic level. In the high-tempo and high-intensity work life, we must not only emphasize the urgency of time, but also try to achieve a quality lifestyle in the intense working hours. And the state, fast food more and more appear in people's daily life, which has great requirements for the production efficiency of food processing[1]. Take the bowled food as an example; the processing flow of a finished product is basically divided into the following processes:Bowling,discharging,tamping, packaging,packing,and storage. There is an important link between the two processes of compaction and encapsulation, that is, the transfer of products from the bowl truck to the packaging line; the existing transfer mode of the enterprise is manual loading. This method is not only labor intensive, but also inefficient,easy to turn over the bowl [2].

2. Overall design

The main task of this design is based on the actual working conditions of the enterprise, and the problems of low efficiency, high strength and easy to turn over the bowl in the working process are the research background. The vacuum packaging line is analyzed under the premise of a certain working period, the manipulator and the vacuum packaging line. The cycle matching problem, according to the scheme for 3D modeling, and based on PLC, touch screen and configuration software to develop and design automatic feeding control system, its main design content: 1. bowl reserveconveyorlinesystem; 2. manipulator follower control system.

IOP Conf. Series: Materials Science and Engineering **626** (2019) 012006 doi:10.1088/1757-899X/626/1/012006

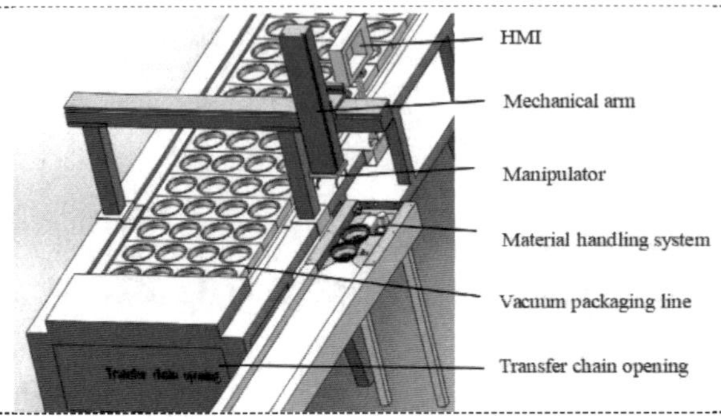

Figure 1. Overallschememodeling.

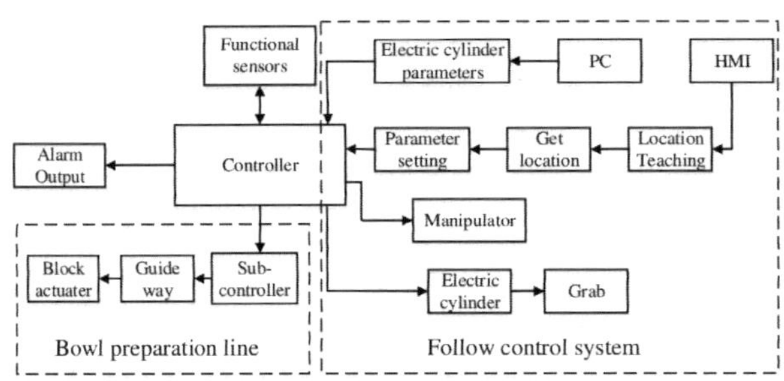

Figure 2. Overallcomposition of the scheme.

2.1. Bowlreserve conveyor line system

The preparation conveying line system comprises two parts: a guiding device and a separating device. The guiding device has the function of bowl guiding to avoid the occurrence of blocking during the feeding process, and the separating device has the function of bowl blocking, and the sensor detects the conveyor belt[3]. Whether the bowl is in place and the signal is transmitted to the steering gear in the separating device, the moving piece is driven to act as a bowl to meet the working requirements of the single grab of the manipulator, and the bowl conveying task of the conveyor belt is completed. Design and process the guide bowl conveying board and the spacer class, map and design the structure and the installation position of each part of the sensor to ensure that the signal can be transmitted to the controller in real time under the premise of accurate operation. Under the control of thesystem, The accurate work of the partition plate part is completed, and the gripper can accurately complete the gripping action when the bowl reaches the position to be grasped[4].

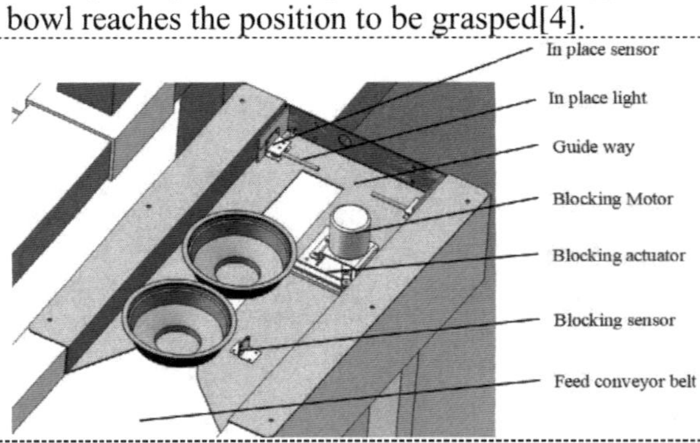

Figure 3. Bowl reserve conveyor line.

During the process of conveying the bowl, the detecting sensor detects that the bowl reaches the position to be grasped, the manipulator moves to grasp, and places the packaging line station; when the detecting sensor and the blocking sensor simultaneously detect the bowl, at this time on the conveyor belt Too much bowl, in order to prevent the accumulation and dumping of bowls during the transfer process, the conveyor belt stops. When the manipulator grabs and transfers the bowl, the conveyor belt continues to run, and the bowl is transferred to the position to be grabbed; the servo sensor detects the bowl entering. When the position working area is to be grabbed, the signal is sent to the steering gear. At this time, the steering gear drives the movement to block the incoming bowl[5]. When the manipulator grabs and transfers the bowl, the movement moves and the bowl continues to be driven to the position to be grasped.

2.2. Manipulator follow control system

The manipulator follows the control system with the process flow as the main line. According to different occasions, it is divided into manual and automatic modes. On this basis, the control circuit is designed, PLC is used as the core controller and the program is developed[6]. The interface of the manipulator's touch screen is designed and the prototype of the manipulator is applied. In the process of clamping and handling, the manipulator is required to move in the positive and negative directions of the X-axis and the positive and negative directions of the Z-axis. Two linear motion units are selected to realize the X-axis and Z-axis movement of the manipulator respectively[7].

- Since the object to be gripped is a plastic bowl of food, the clamped place cannot leave traces, so there is a high requirement for the strength of the gripping. Electric claw with self-adaptive force. The working state of the jaws is: initial state: N0.0 speed 300mm/s, the jaws are opened to 30mm on both sides, the jaws are in the initial state after resetting or system power-on; the state to be grasped: NO.1 speed 300mm /s, both sides of the jaws open to 22mm, waiting for the bowl to reach the specified position to be grabbed; grab state: NO.2 speed 300mm / s, both sides of the jaws contracted to 14mm, the strength is 30%, the bowl is detected When the position to be grasped is reached, the manipulator performs the grasping. When the force reaches 30%, the manipulator lifts up to perform the next placing action.

Table 1. Electric gripper data.

NO	Move	Speed/(mm/s)	Position/mm	Pushing/%
0	Absolute	300	30	30
1	Absolute	300	22	30
2	Absolute	300	14	30

- Accordingto the functional requirements and movement rules of the manipulator, the hardware of the control system is designed, and the PLC controller is programmed in software. The programming of the manipulator is a modular program editing idea[8]. The control system is divided into five parts: system initialization, system parameter setting, working mode selection, linkage operation and error clearing. Before the manipulator is fully automatic linkage operation, the Z axis, the X axis and the electric clamping jaw are reset, and the X axis and the Z axis are linked to each station in the manner of debugging operation, and the position parameter information of the station is obtained. The parameters are written to the PLC program[9].

IOP Conf. Series: Materials Science and Engineering **626** (2019) 012006 doi:10.1088/1757-899X/626/1/012006

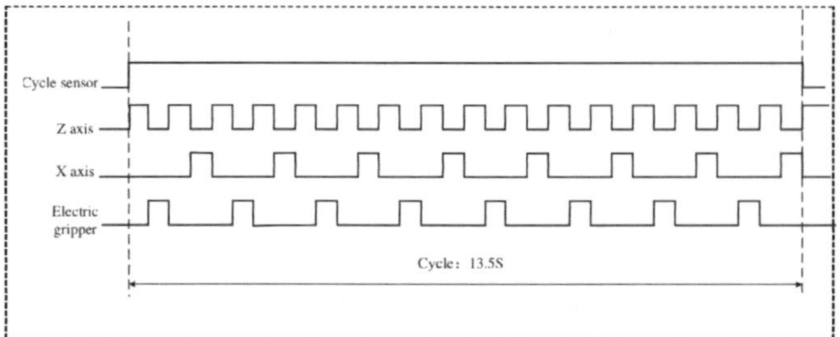

Figure 4. Hardware control system.

3. Conclusion

According to the requirements of the enterprise, a set of mechanical equipment and control system with automatic feeding function was developed, which realized the precise control of automatic feeding of bowled food.1.Due to the need to locate and grab during the feeding process, a mechanical device that can guide and separate bowls is designed to facilitate the grasping of the manipulator; 2. Since the object to be grabbed is a plastic bowl, the force is too large to be deformed, and the force is too small to grasp firmly, so the force feedback manipulator is used for grasping; 3.Select the cantilever type gantry manipulator to carry out the work, and place the bowl accurately at the station to achieve precise position control; 4. The package line is stepped once, and the manipulator works for one cycle to achieve follow control. The whole process is set by the PLC system, and the touch screen is used as the human-computer interaction window, which facilitates data input adjustment and reading, and realizes humanized operation. Improve the efficiency of the production line, reduce the labor force, meet the production cycle, and provide a theoretical basis for the later equipment improvement, with high economic value and engineering significance.

Figure 5. Equipment field work.

References

[1] H aI C . Kinematic parameter calibration method for industrial robot manipulator using the relative position[J]. *Journal of Mechanical Science & Technology*, 2018, 22(6):1084-1090W.

[2] HonarpardazM ,Tarkian M , J. Ölvander, et al. Finger design automation for industrial robot grippers: A review[J]. *Robotics & Autonomous Systems*, 2017, 87:104-119.

[3] Kim J , Muramatsu M , Murata Y , et al. Omnidirectional vision-based ego-poseestimation for an autonomous in-pipe mobile robot[J]. *Advanced Robotics*, 2017, 21(3-4):441-460.

[4] Din Z E , A. S . [IEEE IEEE International Symposium on Industrial Electronics - Warsaw, Poland (17-20 June 1996)] *Proceedings of IEEE International Symposium on Industrial Electronics - High performance* PLC controlled stepper motor in robot manipulator[J]. 2016, 2:974-978.

[5] Bonfè, Marcello, VignaliM ,Fiorini M . PLC-based control of a robot manipulator with closed kinematic chain[C]// *IEEE International Conference on Robotics & Automation*. IEEE, 2009.

[6] Dai F , Becker O . A PLCopen-Based Approach for Utilizing Powerful Industrial Robot Functions in PLC-Controlled Applications[M] *Advances in Sustainable and Competitive Manufacturing Systems*. Springer International Publishing, 2013.
[7] Vinothkumar C, Elavarasan R, Jamunarani D. An integrated virtual control of robot manipulator for automated industrial environment[C]// *International Conference on Circuit*. 2016.
[8] Shalfan A, Khalid A. Development of autonomous navigation wheelchairs based on fuzzy control. *Neural Network World*, 2009: 192):223--233.
[9] Pchelkin S S , Shiriaev A S , Robertsson A , et al. Integrated time-optimal trajectory planning and control design for industrial robot manipulator[C]// *Intelligent Robots and Systems (IROS), 2013 IEEE/RSJ International Conference on. IEEE*, 2013.

IOP Conf. Series: Materials Science and Engineering **626** (2019) 012007 doi:10.1088/1757-899X/626/1/012007

Failure Test Optimization Method of Silicon Pressure Sensor Based on Zero Failure Data

Lei Zuo[1,a], Hongkai Sun [1]

[1] School of Electrical and Automation Engineering, Hefei University of Technology, Hefei 230009, China

benzl0313@126.com

Abstract. With the increasing reliability of the silicon pressure sensor, the phenomenon of zero-failure data often appears in the reliability test. This paper studies the optimal confidence limit method and distribution curve method to obtain the confidence coefficient- and the constant-c(c>1) for the reliability analysis of zero-failure data of silicon pressure sensor, subsequently, the two methods were tested in different groups, by analyzing the influence of the number of test groups on the prediction analysis results, the optimal number of groups for the reliability prediction test of the silicon pressure sensor was obtained. The simulation results show that the best prediction results can be obtained when the confidence coefficient- ∈ [0.3,0.6], c∈[4-6] and the number of groups is 10 to 13. This method improves the accuracy of failure rate estimation and reliability test efficiency with limited samples.

1. Introduction

With the development of technology, the requirement of component reliability is higher and higher. Nowadays, due to the high reliability of components, the traditional reliability evaluation method is no longer applicable to predict their reliability [1]. For the general equipment reliability analysis problem based on Weibull distribution, the least square method can be adopted, assuming that shape parameters remain unchanged and scale parameters change with the characteristics of equipment type, to provide two-parameter Weibull distribution parameter estimation [2].

The characteristic of Bayes method is to make full use of the prior information, but sometimes the prior information is not easy to obtain. For this kind of problem, the prior information can be expressed as the interval estimation of the reliability function [3-4]. In the absence of failure data, the confidence interval of equipment reliability is difficult to estimate. The confidence interval of equipment reliability can be obtained by combining bayes model and variance propagation technique [5], Similarly, for system reliability, a dynamic Bayes estimation method can be used: take the predicted results as the prior knowledge of the reliability in the next stage, and use the maximum entropy method to determine the prior distribution of the system reliability [6]. In order to solve the challenge of fault modeling and prediction caused by zero-failure data, statistical prediction program can be developed in combination with Bayes model to provide the point estimation of the cumulative number of faults in the future period, and the impact of different assumptions on the failure time can also be evaluated [7-10].

Later, regarding the reliability prediction of zero-failure data, some scholars put forward innovative methods such as E-Bayes (Expected Bayesian Method) [11], for the aerobearing whose life obeys Weibull distribution, E-Bayes method is used to study the influence of different parameter values and number of different test groups on the prediction accuracy, and the optimal data for the reliability analysis of aerobearing is finally determined, but the life obeying exponential distribution is not considered [12]. This paper focuses on the study of silicon pressure sensors whose life follows exponential distribution, discuss the parameter estimates of the optimal confidence limit method and the E-Bayes-based distribution curve method to select a suitable method for the analysis of silicon pressure sensors, the appropriate number of test groups can be obtained by testing the above two methods, and the suggestions for reliability evaluation of silicon pressure sensors were given.

2. Related Work

Failure rate is one of the main prediction targets in product reliability analysis. Product life is basically subject to index, normal, weibull distribution, etc. Here, products subject to exponential distribution are discussed and analyzed. About the exponential distribution, the distribution function F(t), the probability density function f(t), reliability function R(t), failure rate function □(t) as shown in the following typeIn this paper, a reliability test optimization method is proposed for the prediction problem based on zero-failure data. The specific implementation steps are given by equation (1), equation (2), equation (3), and equation (4).

$$F(t)=\begin{cases}0, & t\le 0,\\ 1-e^{-\lambda t}, & t>0.\end{cases} \quad (1)$$

$$f(t)=\lambda e^{-\lambda t} \quad (2)$$

$$R(t)=1-F(t)=e^{-\lambda t},\ t>0 \quad (3)$$

$$\lambda(t)=\frac{f(t)}{R(t)}=\lambda \quad (4)$$

The □ is the failure rate of the product, and □ > 0. Figure 1 shows that the probability density curve becomes steeper and steeper with the increase of □.

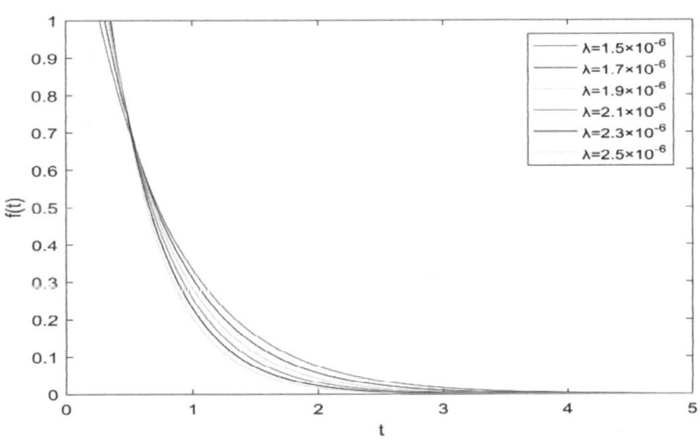

Figure. 1 Exponential probability density

The timing truncation test with a total number of m is carried out for the product, where the truncation time in the ith (i=1, 2……, m) timing truncation test is T_i ($T_1<T_2<\cdots\cdots<T_m$), and the number of corresponding test samples in this group is n_i .If no samples fail at the end of the test, then (n_i , T_i) is the zero-failure data [5].

The design life of silicon pressure sensors in the current market is about 50 years. Based on this, we select 6 groups of medium difference products with failure efficiency □=1.5×10^{-6} to 2.5×10^{-6} for simulation analysis, and obtain the optimal grouping method for life test of silicon pressure sensors within this range. First, Monte Carlo method was used to generate 400 random numbers with failure rates subject to exponential distribution, and the generated random numbers were arranged from small to large in an orderly manner. Every 5 data was a group, and the first 20 Numbers were selected. The smallest number in each group was reserved as a decimal as the timing truncation time-T_i, and it was considered that the product would never fail at the truncation time. The six groups of timing truncation time are shown in Table 1 below.

IOP Conf. Series: Materials Science and Engineering **626** (2019) 012007 doi:10.1088/1757-899X/626/1/012007

Table 1. Timing cut-off time under different failure rates.

$□_1=1.5\times10^{-6}$	$□_1=1.7\times10^{-6}$	$□_1=1.9\times10^{-6}$	$□_1=2.1\times10^{-6}$	$□_1=2.3\times10^{-6}$	$□_1=2.5\times10^{-6}$
613.4	541.2	267.6	1163.2	1371.6	203.4
11659.1	10287.4	5868.6	5780.3	11096.2	4460.1
19898.9	17557.8	16764.5	13164.2	16152.8	12741.0
27567.2	24324.0	30541.6	20961.4	24497.6	23211.6
33052.3	29163.8	36582.7	29636.2	27569.9	27802.8
41958.8	37022.4	46183.4	32636.2	31033.1	35099.4
49202.4	43413.9	56338.9	38500.0	40911.4	42817.6
60405.2	53298.7	62653.4	46816.2	45947.4	47616.6
65963.8	58203.3	73331.5	53931.2	48895.8	55731.9
70200.5	61941.6	79110.9	68097.5	53532.6	60124.3
83809.7	73949.8	88530.4	84272.4	57040.3	67283.1
100981.1	89100.9	95853.1	88933.0	63703.1	72848.3
115666.6	102058.8	104459.9	99402.1	71321.1	79389.5
128752.1	113604.8	118128.4	105926.4	78824.0	89777.6
134232.1	118440.1	127580.9	115619.3	82782.0	96961.5
139181.2	122806.9	132634.4	122918.9	89112.1	100802.2
148528.7	131054.7	138224.5	126367.0	99918.7	105050.6
153542.3	135478.5	143453.9	131876.6	107861.8	109025.0
166451.9	146869.3	154546.5	137543.2	117843.7	117455.3
185418.4	163604.4	160013.7	142966.5	125648.5	121610.4

With the progress of the test, the probability of product failure gradually increases, so it is assumed that n_i decreases successively, and there are s_i products that fail at the time of T_i, The $s_i= n_i+n_{i+1}+......+n_m$, assuming 210 products were tested. Group number and partial truncation time are shown in Table 2.

Table 2. Zero failure data grouping.

Number of clusters	n_i	s_i	Number of clusters	n_i	s_i
1	20	210	11	10	55
2	19	190	12	9	45
3	18	171	13	8	36
4	17	153	14	7	28
5	16	136	15	6	21
6	15	120	16	5	15
7	14	105	17	4	10
8	13	91	18	3	6
9	12	78	19	2	3
10	11	66	20	1	1

Confidence coefficient for the reliability analysis of device zero-failure data is obtained by calculating the classical confidence upper limit $□_{CU}$ and MTBF at different confidence levels using the optimal confidence limit method, given by equation (5), where MTBF, the mean time between failures, is equal to the inverse of the failure rate when the product life obeying the exponential distribution, that is: MTBF=1/□. The formula for calculating the relative error of the failure rate obtained by the classical confidence upper limit $□_{CU}$ is given by equation (6). Then, the optimal grouping number is discussed based on the optimal confidence coefficient-□.

$$\lambda_{CU} = -\ln \alpha / T, MTBF = 1/\lambda_{CU} \tag{5}$$

$$\delta_{CU} = |(\lambda_{CU} - \lambda)/\lambda| \tag{6}$$

When using the distribution curve method, the optimal constant c is obtained first, and then the optimal grouping of the test is discussed. For the product whose life is exponentially distributed, its failure rate □ least squares estimate is given by equation (7).

$$\hat{\lambda} = [-\sum_{i=1}^{m} T_i \ln(1-\hat{p}_i)] / \sum_{i=1}^{m} T_i^2 \tag{7}$$

In equation (7), $\hat{p}_i$ is the E-Bayes estimation of p_i. Assuming that the failure probability of the test product at the time of T_i is p_i, And the multi-layer prior density function □(p_i) is given by equation (8), in the case of square loss, the E-Bayes estimation of p_i is given by equation (9). Where $0< p_i <1$, b is the super parameter, and $1<b<c$ ($c>1$, c is a constant), the prior distribution of b is the uniform distribution on (1, c). The value of c is 2 to 9 [12].

$$\pi(p_i) = \frac{1}{c-1}\int_1^c b(1-p_i)^{b-1}db \tag{8}$$

$$\hat{p}_i = \frac{1}{c-1}\ln(\frac{s_i+c+1}{s_i+2}) \tag{9}$$

3. Failure rate prediction analysis of different groups

Firstly, when using the optimal confidence limit method, the confidence coefficient-□ is changed by using equation (5) and equation (6) under the condition that the number of groups ranges from 1 to 20. The relative error a under different failure rates was calculated, and the results of partial grouping (groups 10, 13 and 15) were shown in the Figure 2.

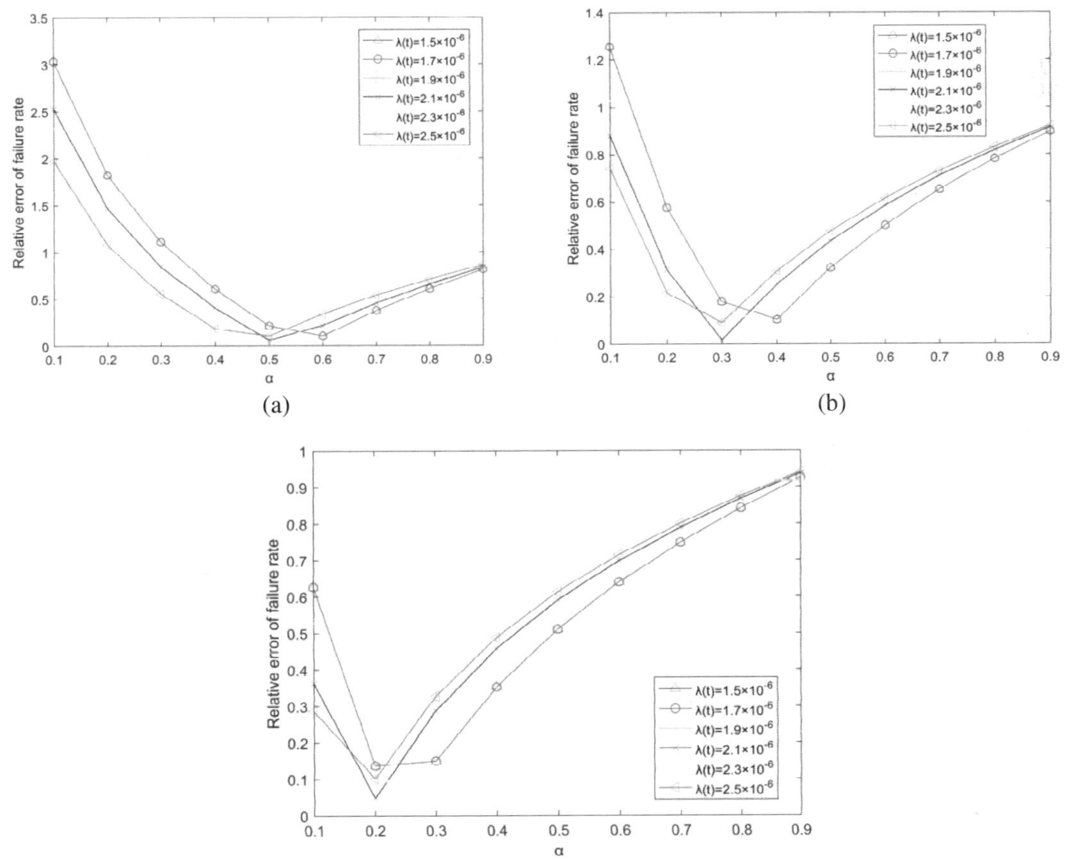

(c)

Figure 2 (a) Relative error of the predicted failure rate with different confidence(X=10). **(b)**Relative error of the predicted failure rate with different confidence(X=13). **(c)**Relative error of the predicted failure rate with different confidence(X=15).

As can be seen from Figure 2, on the whole, the relative error of the predicted value of failure efficiency decreases first and then increases, and the value of optimal confidence coefficient-□$_B$

decreases with the increase of the number of test groups X. And for the same test group, the higher the failure rate is, the smaller the optimal confidence coefficient-$□_B$ is. By comparing Figure 2 (a) with Figure 2 (b), it can be seen that if the number of test groups decreases, the prediction error will increase. At the same time, it can also be seen from Figure 2 (c) that for the optimal confidence limit method, when the confidence coefficient value slightly deviates from the optimal value $□_B$, the predicted value will show a significant deviation. Therefore, we believe that the test data should not be too much, or the phenomenon of "rush" will occur.

Subsequently, on the basis of the optimal confidence coefficient-$□_B$ for each failure efficiency obtained in the above test, the test group was changed from group 1 to group 20 according to equation (5) and equation (6), and the relative error of the failure efficiency-$□_{CU}$ was determined (the relative error of each failure efficiency-$□_i$ in Table 3 is $□_{CUi}$), and the optimal predicted value was obtained as shown in Table 3.

Table 3 The optimal value of each failure rate prediction under different test groups.

	$\alpha_{BX=10}$	$\alpha_{BX=13}$	$\alpha_{BX=15}$
δ_{CU1}	0.0974151	0.0805728	0.0094032
δ_{CU2}	0.0974135	0.0805718	0.0094037
δ_{CU3}	0.1050646	0.07034	0.039844
δ_{CU4}	0.0623902	0.0170986	0.0477619
δ_{CU5}	0.0011957	0.0616408	0.0592041
δ_{CU6}	0.1050637	0.0703409	0.0398447

According to Table 3, generally speaking, with the increase of test samples, the prediction accuracy is higher, with the increase of failure rate, the number of corresponding optimal test groups should also increase. At the same time, on the basis of considering the test cost, the test consumption should be minimized within the allowable engineering error. It can be seen from the above data that when the test groups exceed 10, the relative error is basically controlled within 1%, meeting the engineering requirements. Therefore, in the case of limited test samples, the number of experimental groups can be selected from 10 to 15. The parameter $□_B$ in the optimal confidence limit method is from 0.3 to 0.6, and the larger the failure rate is, the smaller the value of $□_B$ is correspondingly.

Finally, in combination with equation (7) and equation (9), the distribution curve method is used to verify the reliability prediction feasibility of products under the above grouping conditions. In engineering applications, the value of c is determined according to specific products, c here take from 2 to 9. The number of test groups were 10, 13 and 15, respectively. The failure efficiency was estimated by least square method in equation (7), and the results are shown in Figure 3.

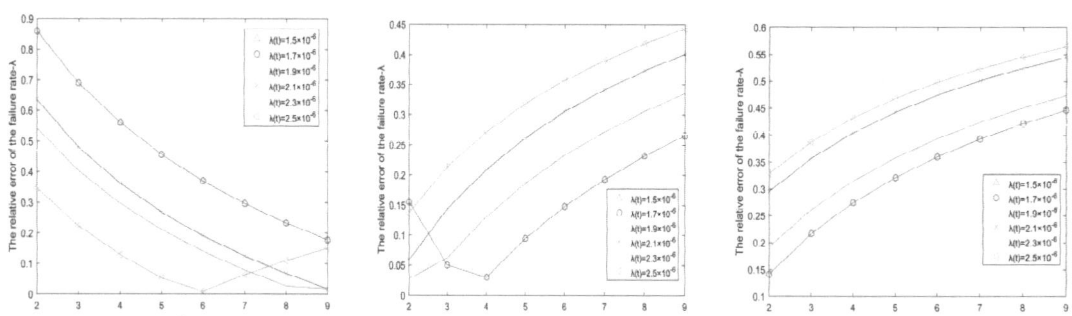

(a) (b) (c)

Figure 3 (a) The least squares estimation of failure rate under different constant c (X=10). **(b)** The least squares estimation of failure rate under different constant c (X=13). **(c)** The least squares estimation of failure rate under different constant c (X=15).

As shown in Figure 3, with the increase of the number of test samples, the constant c applicable to the least squares estimation method of the failure rate □ is decreasing; the constant c suitable for the least squares estimation of the failure rate □ is decreasing with the increase of the failure rate. At the same time, it can be seen that the value range of c that meets the engineering requirements is from 4 to 6. In addition, when the number of experimental groups is 15, the least square estimation is not applicable, so it is suggested that the number of experimental groups should be between 10 and 13.

4. Conclusion

Two widely used methods, optimal confidence limit method and distribution curve method, are used to predict the life of the device. For these two methods, the following Suggestions are made:

1. The optimal number of groups for reliability life prediction test of this type of product can be set as 10 to 13;
2. When using the optimal confidence limit method, for a certain failure efficiency, the relative error presents a nonlinear change, with the increase of confidence, the relative error first decreases and then increases, and the predicted results meet the engineering requirements when is 0.3 to 0.6.At the same time, when the value of □ is less than the optimal confidence coefficient value, the prediction error is getting smaller and smaller with the increase of failure rate □, while when the value of □ is more than the optimal confidence coefficient value, the prediction error shows a trend of getting larger and larger.
3. For the distribution curve method, the failure probability pi in this experiment is estimated by E-Bayes, where the optimal value range of the constant c value is 4 to 6. Under the condition of constant failure efficiency, the optimal value of c gradually decreases as the number of experimental groups increases, and notice that the larger the failure rate, the smaller the value of c.

The life of electronic products generally obeys the exponential or weibull distribution, and the next research plan plans to conduct the prediction analysis of the reliability life of electronic products with no failure data whose life obeys the weibull distribution, and study the optimization method of parameters and group number by considering the influence of environmental factors.

Acknowledgments

This project was supported by Key Projects of Equipment Advance Research (41402040301), National Key Research and Development Program Funding Projects (2016YFF0102200), Key Projects of the National Natural Science Foundation of China (51637004) and National Natural Science Foundation of China (51777050, 51577046).

References

[1] Sanling Song , David W. Coit, Qianmei Feng and Hao Peng, "Reliability Analysis for Multi-Component Systems Subject to Multiple Dependent Competing Failure Processes," *IEEE Trans. Rel.*, 63. 3 (2014): 331-345

[2] Harold S. Balaban and Kent Haspert , "Estimating Weibull Parameters for a General Class of Devices from Limited Failure Data," *IEEE Trans. Rel.*, R-21. 5 (1972): 111-117

[3] M.P. Kaminskiy and V.V. Krivtsov , "A simple procedure for Bayesian estimation of the Weibull distribution," *IEEE Trans. Rel.*, 54. 12 (2005): 612-616

[4] G. Mazzanti, E. Chiodo, G. Mazzanti and E. Chiodo, "Bayesian reliability estimation based on a weibull stress-strength model for aged power system components subjected to voltage surges," *IEEE Trans. Dielectr. Electr. Insul.*, 13. 2 (2006): 146-159

[5] Huairui Guo, Tongdan Jin and Adamantios Mettas, "Designing Reliability Demonstration Tests for One-Shot Systems Under Zero Component Failures," *IEEE Trans. Rel.*, 60. 3 (2011): 286-294

[6] Yunyan Xing, Xiaoyue Wu, Ping Jiang and Qi Liu, "Dynamic Bayesian Evaluation Method for System Reliability Growth Based on In-Time Correction," *IEEE Trans. Rel.*, 59. 6 (2010): 309-312

[7] Jingjing He, Xuefei Guan and Ratneshwar Jha, "Improve the Accuracy of Asymptotic Approximation in Reliability Problems Involving Multimodal Distributions," *IEEE Trans. Rel.*, 65. 12 (2010): 1724-1736

[8] Seung Jun Lee, Sang Hun Lee, Tsong-Lun Chu, Athi Varuttamaseni, Meng Yue, Ming Li, Jaehyun Cho and Hyun Gook Kang, "Bayesian Belief Network Model Quantification Using

Distribution-Based Node Probability and Experienced Data Updates for Software Reliability Assessment," *IEEE Access*., 6. 10 (2018): 64556 - 64568
[9] Zhibing Xu, Yili Hong and William Q. Meeker, "Assessing Risk of a Serious Failure Mode Based on Limited Field Data," *IEEE Trans. Rel.,* 64. 3(2015): 51-62
[10] K.W. Miller, L.J. Morell, R.E. Noonan, S.K. Park, D.M. Nicol, B.W. Murrill, "Estimating the probability of failure when testing reveals no failures," *IEEE Trans. Software Eng.,* 18. 1 (1992): 33-43
[11] V.E. Johnson, A. Moosman and P. Cotter, "A hierarchical model for estimating the early reliability of complex systems," *IEEE Trans. Rel.*, 54. 6 (2005): 224-231
[12] Pandong Gao, Xuejin Shen, Xiaoyang Chen and Tao Zhang, "Reliability analysis for aircraft bearing with zero-failure data," *Journal of Aerospace Power*., 30. 8 (2015): 1980-1987

Two-dimensional Aerodynamic Loads of Thermal Protection System Considering Different Widths of the Strain Isolation Pad

Yupeng Feng, Jialin Cheng, Tao Li, Bin Chen

Technical Center, Chengdu Aircraft Industrial (Group) Co., Ltd, Chengdu 610092, China

feng948837675@163.com

Abstract. Thermal Protection System (TPS) is widely adopted to prevent aircraft substructure from the high environment temperature which is assembled by the ceramic tile, the strain isolation pad (SIP) and the substrate structure. The surface of shuttle is covered by individual ceramic tiles which are bonded to the substrate. The structure of SIP is the significant component of TPS which has important influence on the aerodynamic loads of TPS. To reveal the effect of the width of SIP on the aerodynamic loads, the two-dimensional TPS models considering various widths of SIP, 0.5mm, 0.7mm, 1.0mm, 1.3mm and 1.5mm, are founded and are simulated by computational fluid dynamic (CFD) method. The models are established by combining SIP and the ceramic tile on the surface of NACA0012. The flow around TPS, including flow around NACA0012 and internal flow, is simulated by solving Navier-Stokes (N-S) equation and Spalart-Allmaras (S-A) model. The pressure distributions around TPS are obtained and numerical results show: (1) In X and Y direction, with the width of SIP, $\triangle w$, being 1.0mm, the forces are minimum, 159.9N in X-direction and 313.3N in Y-direction, respectively; (2) The different force of tile-II is related to both the pressure distribution and the flow speed around tile-II.

1. Introduction

The surface of Space Shuttle will produce above 1220F in condition of ascent and re-entry which is the melting point of conventional aluminum aircraft structure [1]. To prevent the aircraft substructure from high temperature, some heat protection material, thermal barrier coating and the insulation structure are pointed out, such as the silica tiles and TPS structure. TPS structure is assembled by the ceramic tile, SIP and the substrate structure is widely adopted to protect the aircraft substructure to avoid the high environment temperature during the ascent and re-entry [2-4]. To protect the aircraft surface from high temperature and avoid the structural damage deriving from the excessive strain, TPS is designed in two ways [5-6]: (1) The aircraft surface is covered by individual ceramic tiles which allow the motion each other; (2) Simple tiles are bonded to the substrate by the deformable SIP. Both internal flows in tile-tile gaps and SIP are produced in this assembly structure. Widths of SIP have significant influence on the porous flow of SIP and the pressure distribution along SIP, which is the important component of the aerodynamic loads of TPS, especially in condition of shock wave on the surface of tile.

Lots of experts have some research on TPS. Lawing [7] obtained the permeability of both high density tile and SIP by performing lots of pressure drop experiments. Muraca [8] pointed out that during

IOP Conf. Series: Materials Science and Engineering **626** (2019) 012008 doi:10.1088/1757-899X/626/1/012008

ascent environments; the biggest aerodynamic loads of tile are happened. Petley [9] calculated the aerodynamic loads of tiles during ascent steady-state by a modified heat-transfer program.
However, the study about the influence of widths of SIP on the aerodynamic loads of tiles is still scarce. In this paper, CFD technology is utilized to calculate aerodynamic loads of TPS considering various widths of SIP, 0.5mm, 0.7mm, 1.0mm, 1.3mm and 1.5mm. The porous flow in SIP is simulated by adopting S-A turbulence model and solving N-S equation.

2. Internal Flow Model

The structure of TPS, which is assembly of the substrate structure, SIP and the ceramic tile, is shown in figure 1 [10]. SIP is utilized to conjunct the simple tile and the substrate. There is internal flows around TPS model, including porous flow of SIP and tile-to-tile gap flow.
To eliminate the coupling effect of adjacent tile, the mechanical model of TPS with three tiles is simplified as the tile-airfoil model in figure 2, which assembles TPS and airfoil NACA0012 together. The shock wave is produced on the upper surface of tiles. The internal flow along with tile-tile gaps and SIP is produced from the high pressure zone on the downstream portion to the low pressure zone on the upstream portion, when the shock waves is on the upper surface of tile. In this paper, the force of TPS is obtained by tile-II which is on the middle tile of TPS.

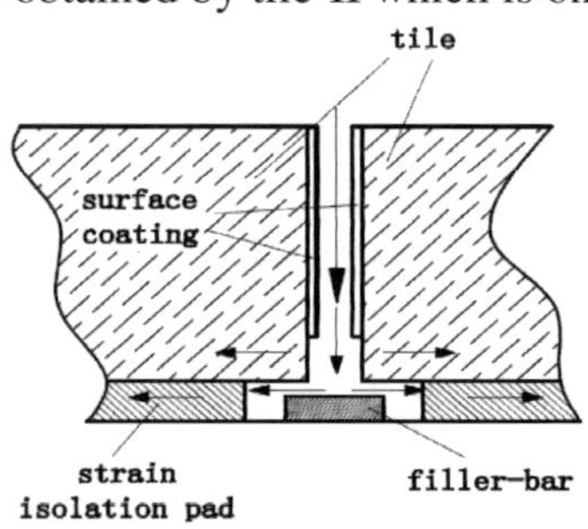

Figure 1. The structure of TPS.

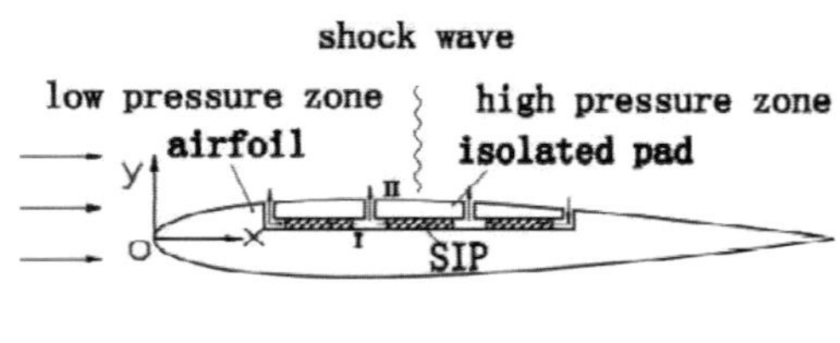

Figure 2. The TPS-airfoil aerodynamic model.

3. Numerical Method

3.1. Governing Equation

Due to solve flows around TPS, the governing equation is as follows in Cartesian coordinates [11]:

$$\frac{\partial \boldsymbol{u}}{\partial t}+\frac{\partial \boldsymbol{f}}{\partial x}+\frac{\partial \boldsymbol{g}}{\partial y}=\frac{\partial \boldsymbol{s}_x}{\partial x}+\frac{\partial \boldsymbol{s}_y}{\partial y} \quad (1)$$

Where $\boldsymbol{f}$ and $\boldsymbol{g}$ are the flow flux in x direction and y direction, respectively; $\boldsymbol{s}_x$ and $\boldsymbol{s}_y$ are the sum of heat exchange flux and viscous stresses in x direction and y direction, respectively; $\boldsymbol{u}$ indicates flow vector.
S-A turbulent model is adopted. What's more, it is chosen that a steady-state pressure-based implicit solver with coupled pressure-velocity and second-order upwind discretization.

3.2. Verification of Algorithm

There are both flow field mentioned above essential to be verified: (1) the porous flow in SIP; (2) the flow around NACA0012. The algorithm has been verified that the numerical method can solve both flow effectively, respectively [10-11].

4. The CFD setup

In this study, the multi-tiles-airfoil model is established. The ceramic tile is about 8.0 cm×1.0 cm and the chord length is 1.0 m. The mesh is all quadrilateral and the grid an O type, in figure 3 [10]. This domain is meshed with 67740 nodes and 68843 elements. Figure 4 shows tile-II in Cartesian coordinates [10]. Boundary conditions (BCs) of the far field are listed in table 1. Additionally, no-slip condition is adopted in this model.

IOP Conf. Series: Materials Science and Engineering **626** (2019) 012008 doi:10.1088/1757-899X/626/1/012008

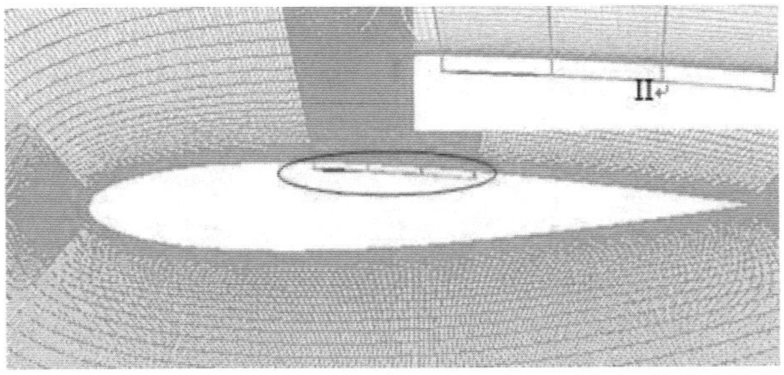

Figure 3. Local mesh of multi-tiles-airfoil model.

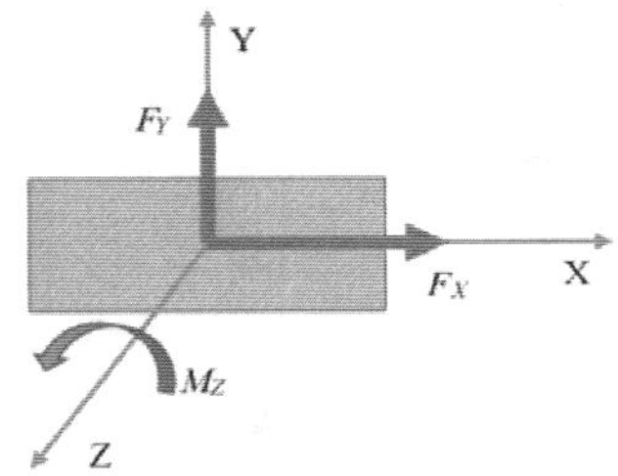

Figure 4. Cartesian coordinates of Tile-II.

Table 1. Boundary conditions

Type	Operating pressure(pa)	Ma	Temperature (K)	Viscosity (kg/m·s)	Permeability of SIP(1/m2)
pressure-far-field	1.6088×10^5	0.8	265.9	1.68×10^{-5}	4.936×10^8

5. Numerical Result

The widths of SIP have influence on the porous flow. Different models with various widths of SIP would be applied to obtain forces of tile-II. Widths of SIP, defined as $\triangle w$, are 0.5mm, 0.7mm, 1.0 mm, 1.3mm and 1.5mm, respectively. The forces of tile-II are shown in table 2. From table 2, it is easily seen that the forces in X and Y-direction are both minimum, when $\triangle w$ is 1.0mm. The force in conditions of $\triangle w$ 1.0mm is chosen to the basis data. As $\triangle w$ is increasing by, the force are increasing 5.4%~ 11.8% and 102.5%~ 127.3% in X-direction and Y-direction, respectively. As $\triangle w$ is decreasing by, the force are increasing 29.7%~ 48.6% and 57.1%~ 93.8% in X-direction and Y-direction, respectively. Compared to condition of $\triangle w$ 1.0mm, the maximum increment force in X-direction is 77.7 N in condition of $\triangle w$ 0.7mm, in Y-direction is 398.9N in condition of $\triangle w$ 1.5mm.

Table 2. The forces in X and Y direction with different widths of SIP.

$\triangle w$(mm)	Fx(N)	ΔFx(%)	Fy(N)	ΔFy(%)
0.5	207.4	29.7	607.2	93.8
0.7	237.6	48.6	492.3	57.1
1.0	159.9	0.0	313.3	0.0
1.3	178.8	11.8	634.5	102.5
1.5	168.6	5.4	712.2	127.3

Notes: The force in conditions of $\triangle w$ 1.0mm is chosen to the basis data.

6. Discussion

It is easily to obtain that the forces of tile-II are related to the width of SIP. When $\varDelta w$ is 1.0mm, the forces in X-direction and in Y-direction are all minimum. There are some factors lead to various forces of tile-II, such as flow field around airfoil and internal flow. The pressure around tile-II is shown in figure 5 and figure 6.

In figure 5, on front surface of tile-II, as $\varDelta w$ is increasing by, the pressure distribution is increasing and the pressure distribution in condition of both $\varDelta w$ 0.7mm and $\varDelta w$ 1.0mm has little difference. On behind surface of tile-II, compared with condition of $\varDelta w$ 0.7mm and $\varDelta w$ 1.3mm, the pressure distribution is less in condition of $\varDelta w$ 1.0mm. As a result, the force in X-direction in condition of $\varDelta w$ 1.0mm is minimum.

In figure 6, on upper surface of tile-II, as $\varDelta w$ is increasing by, the pressure distribution is almost same. Additionally, as $\varDelta w$ is increasing by, the pressure gradient on lower surface of tile-II is diminished.

IOP Conf. Series: Materials Science and Engineering **626** (2019) 012008 doi:10.1088/1757-899X/626/1/012008

What's more, the value of pressure distribution in condition of $\triangle w$ 1.0mm is minimum. As a result, the force in Y-direction in condition of $\triangle w$ 1.0mm is minimum.

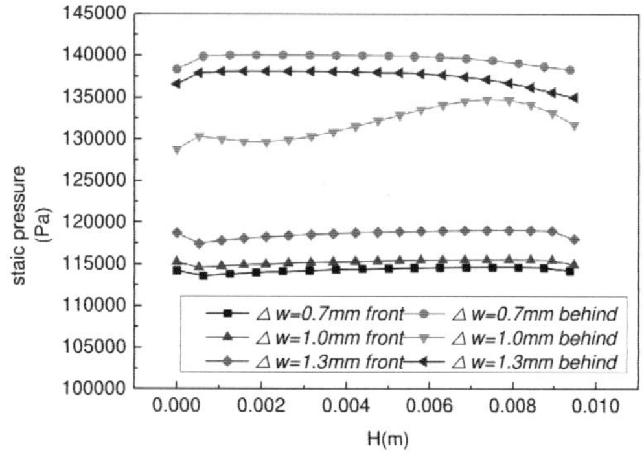

Figure 5. The pressure distribution in X-direction.

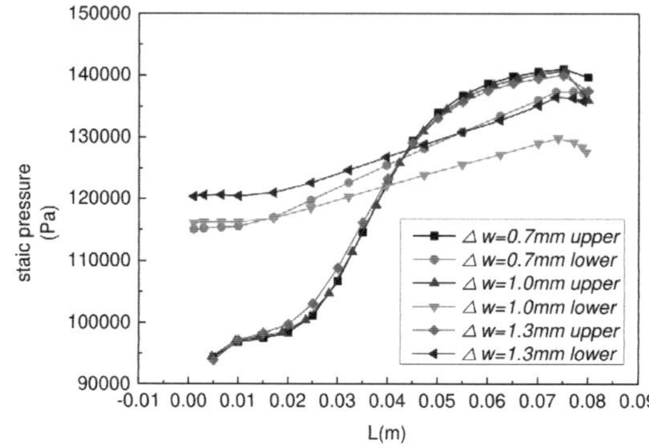

Figure 6. The pressure distribution in Y-direction.

Not only there are various pressure distribution, but various flow speeds. The clouds of local speed in entrance and exit of internal flow are showed in figure 7 and 8, respectively. In condition of $\triangle w$ 1.0mm and $\triangle w$ 1.3mm, the internal flow speed have little difference, average velocity in entrance and exit is about 72.8m/s. Speeds in entrance and exit, in condition of $\triangle w$ 0.7mm, are different from condition of $\triangle w$ 1.0mm and $\triangle w$ 1.3mm, the velocity in entrance is 69.2m/s and in exit is 78.1m/s at table 3. So various pressure distribution and the flow speed around tile-II lead to various force of tile-II.

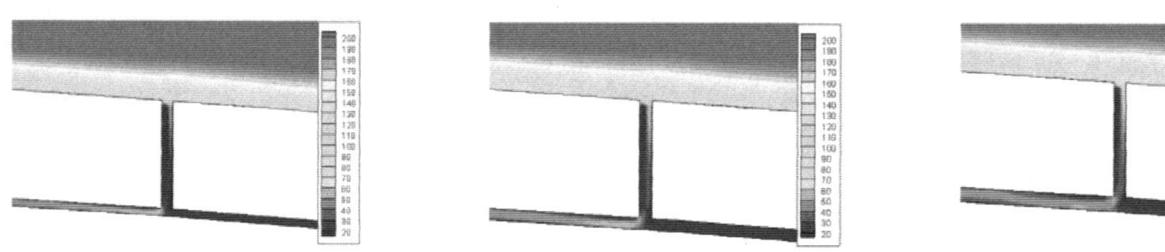

a)0.7mm b)1.0mm c)1.3mm

Figure 7. Clouds of local speed in entrance of internal flow.

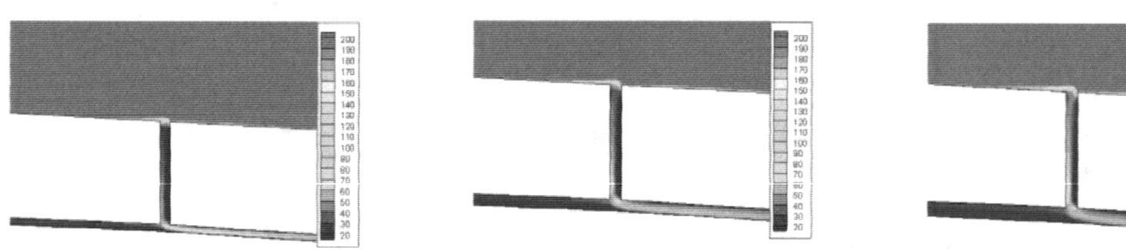

a)0.7mm b)1.0mm c)1.3mm

Figure 8. Clouds of local speed in exit of internal flow.

Table 3. Local flow speed considering various widths of SIP.

$\triangle w$(mm)	The velocity in the entrance(m/s)	The velocity in the exit(m/s)
0.7	69.2	78.1
1.0	72.8	72.9
1.3	72.8	72.6

IOP Conf. Series: Materials Science and Engineering **626** (2019) 012008 doi:10.1088/1757-899X/626/1/012008

7. Conclusion

In this paper, the force around TPS has been studied considering various widths of SIP. The conclusions drawn are:

(1) In X and Y direction, when $\triangle w$ is 1.0mm, the forces are minimum, 159.9N in X-direction and 313.3N in Y-direction, respectively.

(2) The different force of tile-II is related to both the pressure distribution and the flow speed around tile-II.

References

[1] Cooper P A 1981 The shuttle tile story *AIAA*

[2] Rodriguez A C and Snapp C G 2011 Orbiter thermal protection system lessons learned *AIAA* **20** 7308

[3] Monti R, Fumo M D S and Savino R 2006 Thermal shielding of a reentry vehicle by ultra-high-temperature ceramic materials *Journal of Thermophysics and Heat Transfer* **20** 500-6

[4] Wu D, Wang Y, Gao Z and Yang J 2015 Insulation performance of heat-resistant material for high-speed aircraft under thermal environments *Journal of Materials Engineering and Performance* **24** 3373-85

[5] Bertin J J and Cummings R M 2003 Fifty years of hypersonics: where we've been, where we're going *Progress in Aerospace Sciences* **39** 511-36

[6] Blevins R D and Holehouse I 1993 Thermoacoustic loads and fatigue of hypersonic vehicle skin panels *Journal of Aircraft* **30** 971-8

[7] Muraca R J, Coe C F and Tulinius J R 1982 Shuttle tile environments and loads *AIAA* **82** 0631

[8] Lawing P L 1987 A prediction method for flow in the shuttle tile strain isolation pad *AIAA* **87** 1510

[9] Petley D H, Alexander W, Ivey G W and Kerr P A 1984 Steady internal flow and aerodynamic loads analysis of shuttle thermal protection system NASA TP-2255

[10] Feng Y P and Xia W 2019 Two-dimensional aerodynamic loads of Space Shuttle thermal protection system considering steady internal flow *APISAT 2018, LNEE* **459** 151–160

[11] Feng Y P, Xia W, Jiang J S and Hu S L 2016 Transonic aerodynamic loads of airfoil considering internal flow of gaps *FLIGHT DYNAMICS* **34** 15-9 (in Chinese)

IOP Conf. Series: Materials Science and Engineering **626** (2019) 012009 doi:10.1088/1757-899X/626/1/012009

Approximate Design of Optimal Controller for Continuous Disturbance Nonlinear Active Magnetic Bearing System

Peng Liu

Department of Communication Engineering, Chongqing College of Electronic Engineering, 401331, China
pengliu789@126.com

Abstract. A successive approximation algorithm for optimal controller of nonlinear active magnetic bearing system is presented. By constructing a series of linear two-point boundary value problems, the original optimal control problem is transformed into a series of inhomogeneous linear two-point boundary value problems. The optimal control law consists of exact linear term and nonlinear compensation term. The method only needs to solve the nonlinear compensation terms of the adjoint vector differential equation iteratively.

1. Introduction

Active magnetic bearings are widely used in energy, aerospace, high-speed precision machine tools, robots and other fields. Controller is the key component of magnetic bearing technology. The performance of the controller not only determines whether the rotor can be levitated smoothly, but also affects the dynamic performance of the system and the rotating accuracy of the rotor. Magnetic bearings are mainly used as axles or rotors of active suspension systems. Many parts of magnetic bearing have non-linear characteristics, so the whole system is essentially non-linear [1]. In addition to supporting bearings, magnetic bearings can also be used as force actuators and sensors [2]. For example, in rotor dynamics system, Humphris [3] uses magnetic bearings as support to monitor the response of health diagnosis as a method of applying power to the shaft. Similarly, Kasarda et al. [4] proposed a nondestructive evaluation method using magnetic bearing manufacturing process. Gasch [5] model is used to optimize the response of cracked rotor supported by active magnetic bearings. It is found that the vibration characteristics of the system can be changed by introducing breathing cracks, which makes the design and analysis of the magnetic bearing controller very complicated.

The motivation of this work is to develop a successive approximation algorithm to design optimal controllers for active magnetic bearings systems affected by persistent external interference. Based on the state equation of magnetic bearings, the optimal control theory is used to optimize the controller. The difficulty of optimal control for non-linear Active magnetic bearings systems lies in solving the non-linear two-point boundary value problem or the Hamilton-Jacobi-Bellman equation obtained by applying the maximum principle. Actual active magnetic bearing system is affected by external interference [6]. In [7], an internal model structure with adaptive frequency is proposed to eliminate periodic disturbances.

The rest of this paper is organized as follows. The second part establishes the model of the non-linear active magnetic bearing system affected by external interference. In the third part, the design scheme and algorithm of the optimal control law are discussed in detail. The fourth part verifies the validity of the algorithm through simulation examples. See section 5 for conclusions.

2. Problem statement

The motion equation of a nonlinear active magnetic bearing system subjected to continuous external disturbance can be written as a static reference frame

$$x(j+1) = Ax(j) + Bu(j) + f(x(j)) + Dw(j) ,\quad x(0) = x_0 \tag{1}$$

where are the state vector and the input parameter, respectively. Matrix *A, B* and *D* are obtained from equations based on Newton second law, formula of magnetic attractive force and momentum law. x_0 is the known initial vector. $f(x(j))$ is the nonlinear vector. Where $x(j)$and $u(j)$ are state vectors and input parameters respectively. Matrix A, B and D are derived from Newton's second law, magnetic gravity formula and momentum law. x_0 is our initial vector. $f(x(j))$ is a non-linear vector.

Assumption 1. The nonlinear vector $f(x(j))$ satisfies $f: C^1(R^n) \to U \subset R^n, f(0)=0$ and *Lipschitz* condition

$$\|f(z)\| \le \alpha\|z\|,\ z \in U$$
$$\|f(z_2) - f(z_1))\| \le \beta\|z_2 - z_1\|,\ \ z_1, z_2 \in U \tag{2}$$

where α and β are some positive constants.

Assumption 2. The dynamic characteristics of external disturbance vector w satisfies

$$w(j+1) = Gw(j) \tag{3}$$

where *G* is constant matrix. In the following consideration, we assume that

(a) All eigenvalues $\lambda_i (i = 1,2,\cdots,m)$ of matrix *G* satisfy

$$|\lambda_i| \le 1$$

(b) Initial value $w(0)$ of exosystem (3) is unknown.

The quadratic cost functional is

$$J = \frac{1}{2}x^T(N)Q_f x(N) + \frac{1}{2}\sum_{k=0}^{N-1}[x^T(j)Qx(j) + u^T(j)Ru(j)] \tag{4}$$

where $Q, Q_f \in R^{n\times n}$ are the positive semi-definite matrices, $R \in R^{r\times r}$ the positive definite matrix. The optimal control problem is to find a control law u^*, such that the quadratic cost functional (4) is minimized subject to systems (1).

Applying the maximal principle to the Active magnetic bearings systems, together with (4), we have

$$u(k) = -R^{\ 1}B^T\lambda(j+1) \tag{5}$$

where $\lambda(k+1)$ is a solution of the nonlinear two-point boundary value problem

$$\lambda(j) = Qx(j) + (A^T + \frac{\partial f^T(x(j))}{\partial x(j)})\lambda(j+1)$$
$$x(j+1) = Ax(j) + f(x(j)) - BR^{-1}B^T\lambda(j+1) + Dw(j) \tag{6}$$
$$\lambda(N) = Q_f x(N), \qquad x(0) = x_0$$

which is the optimality necessary condition.

Consider nonlinear systems described by

$$z(j+1) = G(j)z(j) + h(z(j))$$
$$z(0) = z_0 \tag{7}$$

where z is the state vector, z_0 is the initial state vector. Suppose that $h(z)$ satisfies the Lipschitz conditions.

Lemma 1. Define vector function sequence $\{z^{(j)}\}$ as

IOP Conf. Series: Materials Science and Engineering **626** (2019) 012009 doi:10.1088/1757-899X/626/1/012009

$$z^{(0)}(j) = \prod_{m=1}^{k} G(j-m)z_0,\ z^{(k)}(0) = z_0,$$
$$z^{(k)}(j) = \prod_{m=1}^{k} G(j-m)z_0 + \sum_{i=0}^{k-1}\left\{[\prod_{m=1}^{k-i-1} G(k-m)]h(z^{(k-1)}(i))\right\}, \tag{8}$$
$$j = 1,2,\dots,\ k = 1,2,\cdots,N,$$

where $G(j-m)$ is the state-transition matrix, and $\prod_{m=1}^{0} G(j-m) = I$ .Then the sequence $\{z^{(k)}\}$ uniformly converges to the solution of the systems (7).

Proof. From (8), we have

$$\left\|z^{(1)}(j) - z^{(0)}(j)\right\| = \left\|\sum_{i=0}^{k-1}\left\{[\prod_{m=1}^{k-i-1} G(j-m)]h(\mathbf{z}^{(j-1)}(i))\right\}\right\| \tag{9}$$
$$\le \alpha\left\|z^{(0)}(j)\right\|\left\|\sum_{i=0}^{k-1}\left[\prod_{m=1}^{k-i-1} G(j-m)\right]\right\| \le \alpha b^2 cj$$

where $\|\cdot\|$ denotes any appropriate vector or matrix norm, and

$$b = \max\{1, \bar{b}\},\quad c = \|z_0\|$$
$$\bar{b} = \sup_{k\in[0,N)}\left\{\left\|\prod_{m=1}^{k-i-1} G(j-m)\right\|\right\} \tag{10}$$

Similarly,

$$\left\|z^{(2)}(j) - z^{(1)}(j)\right\| \le \alpha b^3 c\beta \frac{j^2}{2!}, \tag{11}$$

Further, we derive

$$\left\|z^{(k)}(j) - z^{(k-1)}(j)\right\| \le \alpha b^{j+1}\beta^{j-1}c \frac{j^k}{k!}, \tag{12}$$

According to trigonometry inequality, for any positive integer p,

$$\| z^{(k+p)}(j) - z^{(k-1)}(j) \| \le bc\sum_{i=k}^{k+p} \frac{\beta^{i-1}(bj)^i}{i!}, \tag{13}$$

Hence,

$$\lim_{j\to\infty}\left\|z^{(j+p)}(k) - z^{(j-1)}(k)\right\| = 0. \tag{14}$$

Thus, when k is fixed, $\{z^{(j)}\}$ is a Cauchy sequence, it follows that the sequence is uniformly convergent to the solution of (7) in the form of

$$z(k) = \prod_{m=1}^{k} G(j-m)\varphi(0) + \sum_{i=0}^{k-1}\left\{[\prod_{m=1}^{j-i-1} G(k-m)]h(z(i))\right\}, \tag{15}$$
$$z(0) = z_0$$

The proof is complete

3. Main result

Let

$$\lambda(j) = P(j)x(j) + \bar{P}(j)w(j) + g(j) \tag{16}$$

where $P(k)$ is the unique positive-definite matrix solution of the Riccati matrix difference equation

$$P(j) = Q + E(j+1)P(j+1)A$$
$$P(N) = Q_f \tag{17}$$

IOP Conf. Series: Materials Science and Engineering **626** (2019) 012009 doi:10.1088/1757-899X/626/1/012009

where $E(k+1) = A^T[I - P(k+1)BS^{-1}(k+1)B^T]$ and $S(k+1) = R + B^T P(k+1)B$. $\bar{P}(j)$ is the unique solution to the matrix difference equation

$$\begin{aligned} \bar{P}(j) &= E(j+1)\left[\bar{P}(j+1)G + P(j+1)D\right] \\ \bar{P}(N) &= 0 \end{aligned} \tag{18}$$

$g(k)$ is an adjoint vector to be solved. From (5), (6), (16), (17) and (18), we have

$$\begin{aligned} g(j) &= E(j+1)[P(j+1)f(x) + g(j+1)] + F\lambda(j+1) \\ g(N) &= 0 \end{aligned} \tag{19}$$

$$\begin{aligned} x(j+1) &= [I + BR^{-1}B^T P(j+1)]^{-1}[Ax(j) + f(x) \\ &\quad + [D - BR^{-1}B^T\bar{P}(j+1)G]w(j) - BR^{-1}B^T g(j+1)], \\ x(j) &= x_0 \end{aligned} \tag{20}$$

where $F = \partial f^T(x)/\partial x$. u^* denotes the optimal control which can be expressed as

$$\begin{aligned} u^*(k) &= -S^{-1}(j+1)B^T P(j+1)[Ax(j) + f(x(j)) + Dw(j)] \\ &\quad - S^{-1}(j+1)B^T\bar{P}(j+1)Gw(j) - S^{-1}(j+1)B^T \lim_{j\to\infty} g^{(j)}(j+1), \end{aligned} \tag{21}$$

In order to solve the two-point boundary value problem (6), we construct the adjoint vector difference equation sequence $\{g^{(j)}(j)\}$ as

$$\begin{aligned} g^{(0)}(j) &= g^{(1)}(j) = 0, \\ g^{(j)}(j) &= E(j+1)[P(j+1)f(x^{(k-1)}(j)) + g^{(k)}(j+1)] \\ &\quad + F^{(k-1)}\lambda^{(k-1)}(j+1) \\ g^{(k)}(N) &= 0 \end{aligned} \tag{22}$$

where $F^{(k)} = \partial f^T(x^{(k)}(j))/\partial x^{(k)}(j)$. The state vector difference equation sequence $\{x^{(k)}(j)\}$

$$\begin{aligned} x^{(0)}(j) &= 0, \\ x^{(k-1)}(j+1) &= [I + BR^{-1}B^T P(j+1)]^{-1}[Ax^{(k-1)}(j) + f(x^{(k-2)}(j)) \\ &\quad + [D - BR^{-1}B^T\bar{P}(j+1)G]w(j) - BR^{-1}B^T g^{(k-1)}(j+1)], \\ x^{(k-1)}(0) &= x_0 \end{aligned} \tag{23}$$

And the corresponding optimal control sequence $\{u^{(k)}(j)\}$

$$\begin{aligned} u^{(k)}(k) &= -S^{-1}(j+1)B^T P(j+1)[Ax^{(k)}(j) + f(x^{(k-1)}(j)) + Dw(j)] \\ &\quad - S^{-1}(j+1)B^T\bar{P}(j+1)Gw(j) - S^{-1}(j+1)B^T g^{(k)}(j+1) \\ u^{(0)}(j) &= 0 \end{aligned} \tag{24}$$

For the jth optimal problem, optimal state trajectory and optimal control law are $x^{(k)}$ and $u^{(k)}$, respectively.

4. Simulation

Consider a single input 2-order nonlinear discrete-time system with disturbances described by (1) and (3), where

$$x(k) = \begin{bmatrix} x_1(k) \\ x_2(k) \end{bmatrix}, \quad A = \begin{bmatrix} 1 & 0.1 \\ 0.02 & 1.05 \end{bmatrix}, \quad B = \begin{bmatrix} 0 \\ 0.3 \end{bmatrix}, D = \begin{bmatrix} 0.95 & -0.05 \\ 0.1 & 1 \end{bmatrix},$$
$$f(x) = \begin{bmatrix} 0 \\ 0.25/(1 + x_1^2(k)) \end{bmatrix}, \quad x(0) = \begin{bmatrix} 0 & 0 \end{bmatrix}^T \tag{25}$$

In the following, for the cases of the exosystem (3) are asymptotically stable and stable but not asymptotically stable, we will give the iterative algorithm results, respectively.

Let $G=\begin{bmatrix} 0 & -1 \\ 0.9344 & 1.76 \end{bmatrix}$, $w(0)=[-1.8 \quad 0.4]^T$. It is easy to testify that $\lambda_{1,2}=0.88\pm 0.4i$. The exosystem (3) is asymptotically stable because of $|\lambda_{1,2}|<1$.

Let

$$J = \sum_{k=0}^{\infty}[2x_1^2(j) + u^2(j)] \tag{26}$$

Adopting 6th-order suboptimal control law, the simulation curves of x_1 and u are showed in Fig.1.

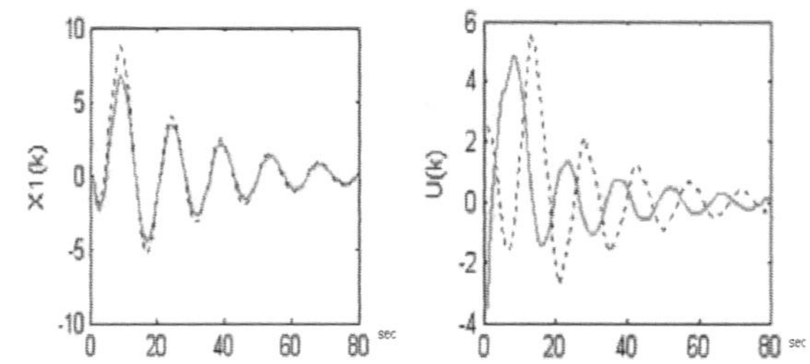

------6th-order FSC ——6th-order FFSC

Figure. 1 Optimal control comparative curves when Exosystem (3) is asymptotically stable

From Fig. 1, we can see clearly that the control effect of FFSC is much better than that of the FSC, and the former is more robust with respect to external persistent disturbances than that of the latter. This example also demonstrates that the higher the iterative times are, the better the control quality is. In addition, the more iterative steps, the higher the control precision. It is important to notice that in the proposed algorithm only a few iteration steps are required in order to get the suboptimal control law.

5. Conclusion

An algorithm for the optimal control of the nonlinear Active magnetic bearings systems affected by external persistent disturbances has been presented derives from SAA to the difference equation in this paper. For the same system, the more iteration times of the nonlinear compensating term in state feedback control law, the better the control effect.

6. References

[1] J.C. Ji and C.H. Hansen, "Nonlinear oscillations of a rotor in active magnetic bearings," *Journal of Sound and Vibration*, vol. 240, no. 4, pp. 599 - 612,2011.

[2] G. Mania, D.D. Quinna and M. Kasarda, "Active health monitoring in a rotating cracked shaft using active magnetic bearings as force actuators," *Journal of Sound and Vibration,* vol. 294, pp. 454-465, 2016.

[3] R. R. Humphris, "A device for generating diagnostic information for rotating machinery using magnetic bearings," in: *Proceedings of the Magnetic Bearings, Magnetic Drives, and Dry Gas Seals Conference and Exhibition*, Alexandria, VA, pp. 123 - 135, 1992.

[4] M. Kasarda, J. Imlach and P.A. Balaji, "The concurrent use of magnetic bearings for rotor support and force sensing for the nondestructive evaluation of manufacturing processes," in: *SPI Seventh International Symposium on Smart Structures and Materials*, Newport Beach, CA, 2010.

[5] R. Gasch, "A survey of the dynamic behavior of a simple rotating shaft with a transverse crack," *Journal of Sound and Vibration,* vol. 160, no.2, 313 - 332, 1993.

[6] H. Q. Zhu, "The study and application on digital control of magnetic bearing," Nanjing: Nanjing University of Aeronautics & Astronautics, 2010 (in Chinese).

[7] R. Marino, G. L. Santosuosso and P. Tomei, "Robust adaptive compensation of biased sinusoidal disturbances with unknown frequency," *Automatica*, vol. 39, no. 10, pp. 1755–1761, 2013.

IOP Conf. Series: Materials Science and Engineering **626** (2019) 012010 doi:10.1088/1757-899X/626/1/012010

Simulation of Stress Distribution and Deformation Analysis for Cantilever I-beams Using ABAQUS Software

Wen Runsheng[1], **Zhang Yang**[1 *]

[1] Jiangsu Province Xuzhou Technician Institute, Xuzhou, Jiangsu Province, China

113503287@qq.com

Abstract. In this paper, the stress distribution of the I-beam cantilever beam is simulated by ABAQUS software, and the uniform load increases with time on the surface of the I-beam to observe the position of the maximum stress and the deformation of the right end.

1. Introduction

Finite element simulation design is a popular design method in recent years. It has the advantages of being intuitive, reliable, fast, and high precision, and has received more and more attention in various industry fields[1]. The basic principle of the finite element method can be summarized as "normalization is zero, set zero is integer", that is, the whole is first divided into a set consisting of a finite number of units, and the units in the set can only be connected by a limited number of nodes. Boundary conditions can be added to these connected nodes instead of the actual overall applied load; then, according to the concept of block approximation, each unit is listed with an equation representing the relationship between the interaction force of the node and the target quantity; finally, according to certain conditions or principles The equations are summarized and the boundary conditions are applied. A finite number of numerical quantities can be obtained by solving. This finite number of numerical solutions can be used to approximate the infinite unknowns in the actual situation[2].

The ABAQUS finite element simulation software is particularly powerful, it not only solves the linear problem, but also solves many difficult nonlinear problems. ABAQUS can analyze: static stress, static displacement, viscoelastic response, viscoplastic response, heat conduction, mass diffusion, coupling, nonlinear dynamic stress, nonlinear dynamic displacement, transient temperature coupling, transient displacement coupling, quasi-static Wait[3].

In this paper, the model of Q235 I-beam is established by ABAQUS software, and the uniform distribution load with time is applied to the upper surface. To observe the stress distribution and deformation. Then continuously increase the load until it exceeds its yield limit, causing the work piece to fail.

2. Finite element simulation of I-beam

Select the standard H-beam as the research object, the section height × width is 100mm × 100mm, the web thickness is 6mm, the flange thickness is 8mm, the process fillet is 10mm, and the beam length is 1m, the material is Q235. The model and meshing established by ABAQUS software are shown in Fig.1. Using finite element software to analyze the model, it is very important to divide the grids. The meshing is too rough to affect the accuracy of the simulation results. If the meshing is too roughly affecting the accuracy of the simulation results, if the meshing is too fine require high computer hardware and affect the calculation speed. This model is simpler and is static analysis, so the mesh can be drawn more carefully, and can close to the actual situation.

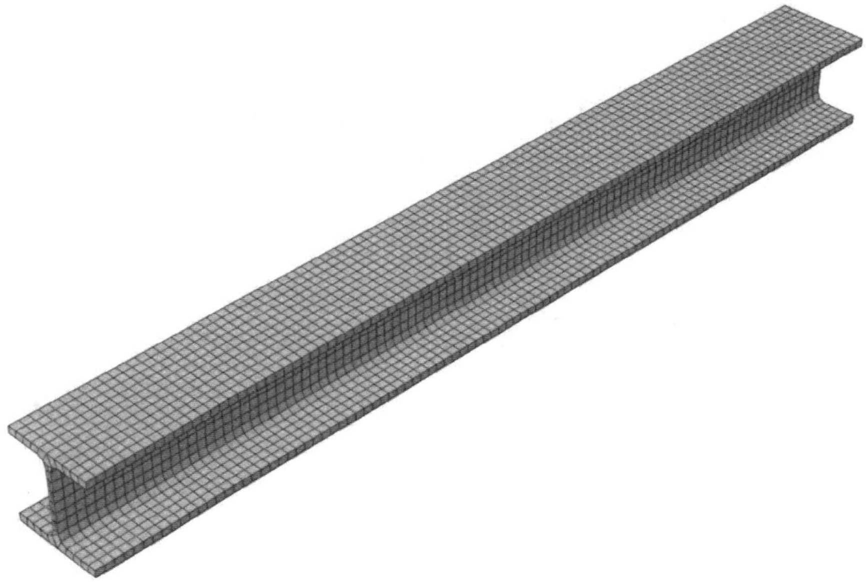

Figure1. I-beam model

Fix the left end of the beam, applying a uniform load that increases over time on the upper surface of the beam. It can be seen from the simulation results that the stress change of the I-beam is gradually increased from left to right and the stress on the flange is greater than that on the web. Then set the yield limit to 235MPa(Yield strength of Q235 steel). As the load increases, the part of the material will fail. The failure will gradually extend upward and downward from the lower left corner(Fig.2). The failure here is not the true failure of the material, but the stress value reaches the yield limit of the material. That is to say, as the load on the upper surface increases, the stress at the two sharp corners of the lower left section is the largest, and the yield limit is first reached, but the material does not fail because the compressive strength of the steel is much greater than its tensile strength. The upper end of the I-beam is pulled, the lower end is pressed, and the upper end is first destroyed.

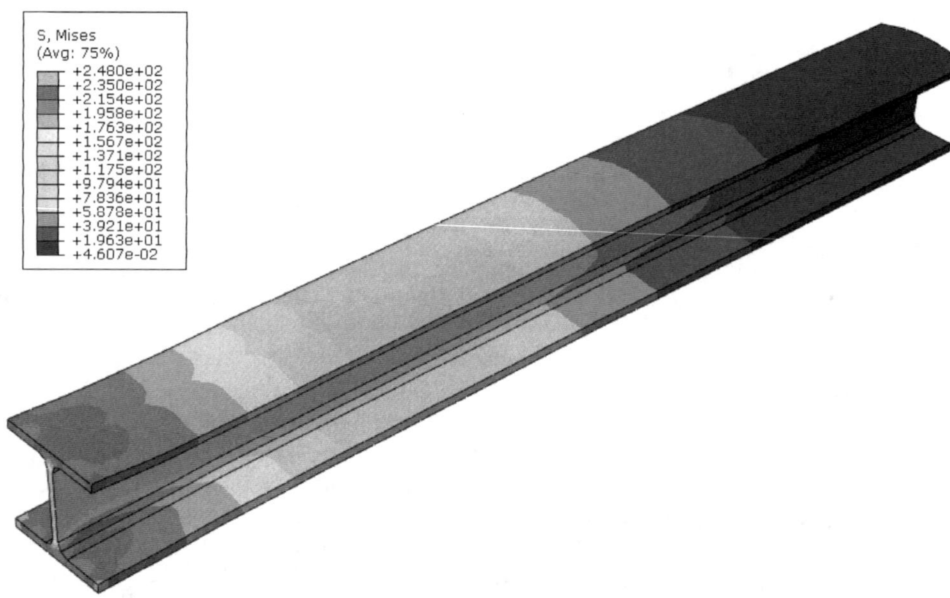

Figure 2. Stress distribution cloud

The engineering strain is the ratio of the difference between the instantaneous gauge length of the specimen and the original gauge length to the original gauge length, it can be expressed as follows:

$$e = \Delta L / L_0 \quad (1)$$

where L_0 is initial gauge length of the specimen; ΔL is increment of L_0.

In fact, once the load is applied, the material is longitudinally elongated, and the cross-sectional size is reduced due to the Poisson effect. The true stress should be equal to the axial force divided by the current cross-sectional area A (instead of the original area A_0). It can be expressed as follows:

$$\sigma = F/A = F/A_0 \bullet A_0/A = S \bullet L/L_0 = S\,(1+e) \quad (2)$$

For any load increment dF, the strain increment $d\varepsilon$ is equal to the ratio of the length increment dL to the current length L (not the original length L_0). Therefore, the true strain ε should be defined as:

$$\varepsilon = \int_{L_0}^{L} \frac{dL}{L} = \ln\left(\frac{L}{L_0}\right) = \ln\frac{L_0+\Delta L}{L_0} = \ln\,(1+e) \quad (3)$$

From the above analysis, the conversion relationship between the engineering stress-strain curve and the real stress-strain curve is as follows:

$$\begin{cases} \sigma = S\,(1+e) \\ \varepsilon = \ln\,(1+e) \end{cases} \quad (4)$$

From the formulas (1)-(4)[4-8], the stress-strain results of the model simulation for small deformation are more accurate, and the model simulation results for large deformation are quite different from the actual situation, or the actual measurement transformation is required to obtain the real result. The I-beam model in this paper is a small deformation model, so the results obtained are relatively accurate. The curve of the displacement of the right end downward with time is shown in Fig. 3. The amount of displacement is small, and the amount of deformation increases with time. Maximum increase to 1.5×10^{-4}mm.

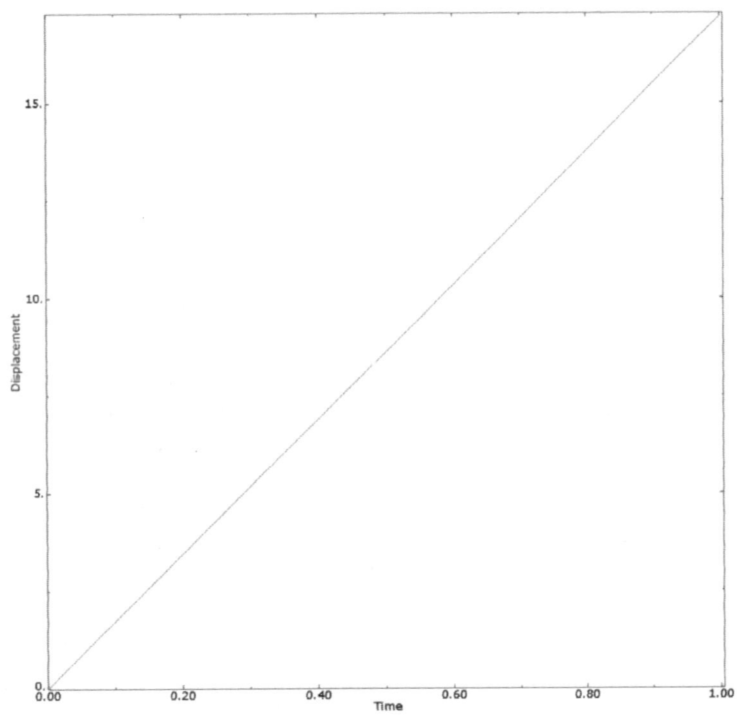

Figure 3. Displacement-time curve

Figure 4 shows the curve of strain energy with time. Strain energy is the potential energy stored in an object in the form of strain and stress, also known as deformation energy. Elastic strain energy means that during the deformation process, the work done by the external force is converted into energy stored in the solid. Under the action of external force, the stored energy is called deformation energy or strain energy. The deformation energy has elastic deformation energy and plastic deformation energy. When the external force is gradually reduced and the deformation is gradually reduced, the solid will release part of the energy and work, and this part of the energy is elastic deformation energy. That is, as the loading force continues to increase, the amount of compression deformation of the I-beam is continuously increased, and the deformation energy stored in the beam is also continuously increased.

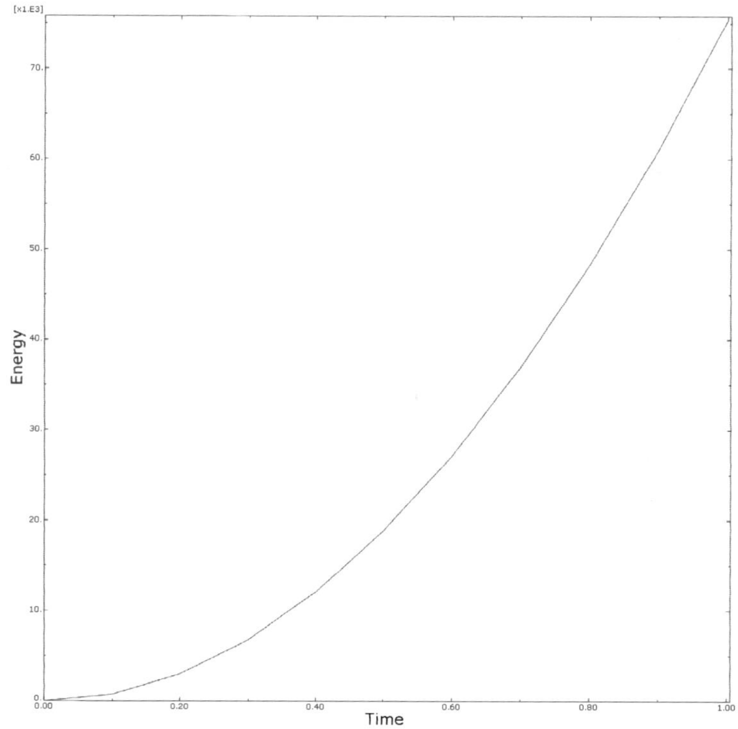

Figure 4. Stress energy-time curve

3. Conclusion

In this paper, the load increased with time on the I-beam cantilever beam, and the ABAQUS software is used to simulate the stress change, and the deformation and the maximum stress position are obtained. It provides a convenient method for optimizing the structural material and examining the applicability of the structural material. By continuously increasing the surface load and observing the deformation amount and stress energy of the beam, the maximum deformation capacity that satisfies the practicability is set to obtain the maximum bearing capacity, and the part of the I-beam that is easy to be broken can be found according to the model, and is appropriately reinforced at the weak point. This is the advantage of software simulation, the software is difficult to detect the workpiece simulation, pre-judgment of the material can be used, the initial detection failure or the choice of reinforcement.

4. References

[1] TAN Jian. Application of Finite Element to Practice Engineering[J]. Chinese elevator.2018,12. 23(29)

[2] YU Yunyun. Contact analysis of split rolling-sliding bearing based on ABAQUS [D]. East China Jiao tong University. 2018

[3] Zhang Linwei. The Research on the Seismic Behavior of T Shaped Side Joints Strengthened with Post Anchorage Tendons[D]. Ji Lin Jian Zhu University. 2018

[4] DENG Xiao-tao， PI Zi-zi， ZHANG Zhi-yi， DENG Yi-tai. The Residual Deformation Analysis of the Cooling Plate of an Aeroengine[J]. Machinery Design & Manufacture,2019.1.1 (55-57)
[5] Zhao Shao-bian.Designing Against Fatigue-method and Date[M]. Beijing： China Machine Press， 1997.
[6] Chen Chuan-yao.Fatigue and Fracture[M]. Wuhan: Huazhong University of Science and Technology Press， 2002.
[7] Wang Xu-cheng.Finite Element Method[M]. Beijing: Tsinghua University Press， 2003.
[8] Ted Belytschko, Wing KamLiu, BrianMoran. Nonlinear Finite Elements for Continua and Structures[M]. John Wiley & sons Ltd, 2000.

IOP Conf. Series: Materials Science and Engineering **626** (2019) 012011 doi:10.1088/1757-899X/626/1/012011

The Quality Control of Oil Leakage in Weld Seam of Small Head Rod of a Common Piston Rod

Zhi-ling Wang

Jiangsu Province Xuzhou Technician Institute, Xuzhou 221151, China

379713268@qq.com

Abstract. We study mainly the reasons of shallow insertion depth of positioning sleeve of piston rod, welding sequence, welding parameters and small foot size. Then, we take the following measures, such as improving the welding sequence and position, strictly controlling welding process parameters, improving the quality awareness of welding operators and controlling the depth of positioning sleeve. Finally, we have perfectly controlled the quality problem of plugging and welding oil leakage of the small end rod.

1 Introduction

The 21st century is a century of quality. With the deepening of China's globalization process, various products have participated in international market competition, and product quality has become the main problem for domestic products to further expand the market and the enter high-end [1].

The hydraulic cylinder is an actuator commonly used in hydraulic system [2]. It is a device that converts liquid pressure into mechanical energy. The pressure relief caused by oil leakage of hydraulic cylinder is one of the main reasons that make the hydraulic cylinder lose its function. The products described in this paper are used for crane telescopic boom [3].

Because the telescopic boom of crane produced now has no self-locking function, once oil leak occurs, it will lead to rapid retraction of telescopic boom, and the result is machine destruction. Most of the oil leakage of hydraulic cylinders is caused by the welded joints, so it is urgent to solve the problem of oil leakage caused by welds [4].

2 Structure of QY12.16II Piston Rod

This paper takes the QY12.16II piston rod of a large crane manufactured by a factory in Xuzhou City, China as an example. The piston rod is mainly composed of 45 # steel rod body and core pipe (as shown in Figure1). The core pipe is welded by 45 # steel rod plug, 20 # steel tubing and 45 # steel earrings.

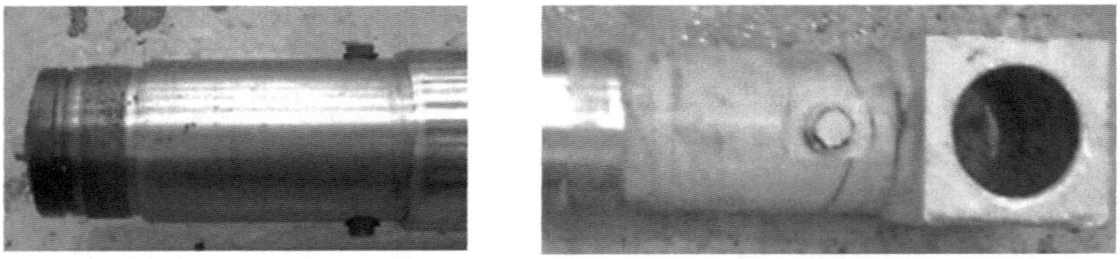

Figure 1. The appearance structure of QY12.16II piston rod

The welding equipment used on the small end rod plug is Panasonic KR500 gas shielded welding machine with 10-meter derrick rubber pulley. The welding wire used is ER50-6, and the welding position is the weld path composed of rod body, rod plug and positioning sleeve . The leakage location is shown in Figure 2.

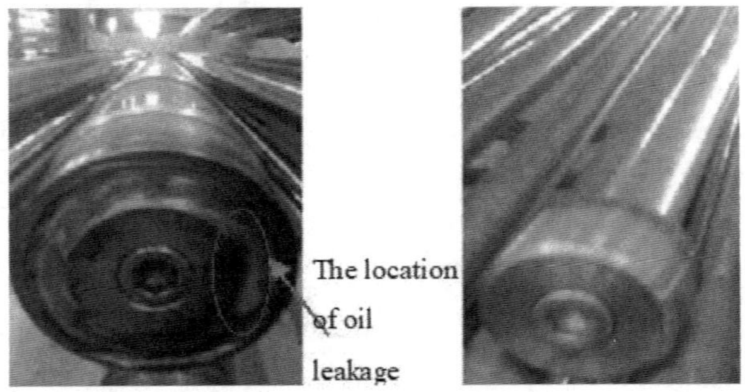

Figure 2. Oil Leakage Location

3 Analysis of the Main Causes of the Problem

The length of QY12.16II piston rod is 7 meters. Although three supporting plates are added to the core tube (as shown in Figure 3), it is unavoidable that the welding seam with one end of the positioning sleeve will be cracked due to the restraint of rod diameter and the influence of gravity [5]. At the same time, due to the unreasonable welding sequence of operators and the factors such as operation without drawing process and welding defects, the oil leakage in the weakest part of the product. the oil leakage in the welding zone, was detected at the pressure test station.

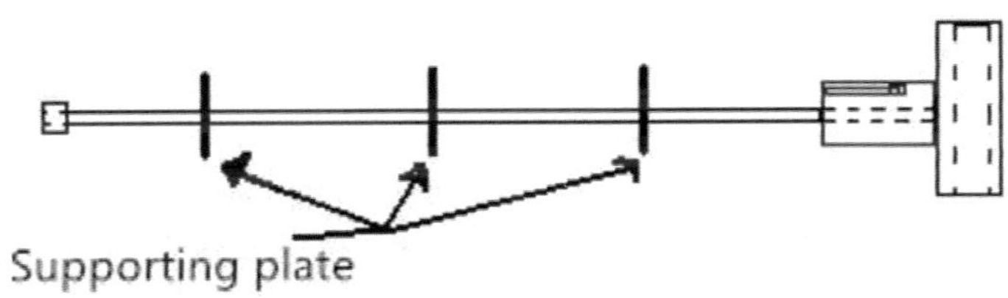

Figure 3. Distribution of the supporting plate position

After in-depth analysis, the specific causes of oil leakage in welds are as follows:

3.1 Improper insertion depth of positioning sleeve

The operator has great randomness in the process of assembling the positioning sleeve for the drawing does not specify the depth of the positioning sleeve, the QY12.16II piston rod does not need to determine the coaxiality between the rod plug and the rod body and the piston is directly mounted on the rod body.As shown in Figure 4, many operators just put the positioning sleeve in 3-5 mm to catch up with the production schedule and began to weld[6].

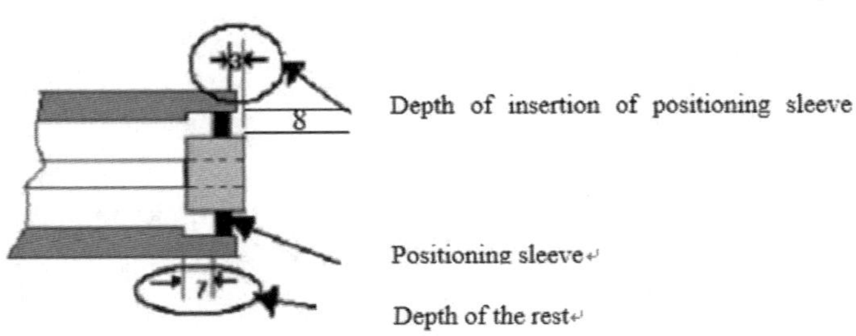

Figure 4. Depth of placement of positioning sleeve

IOP Conf. Series: Materials Science and Engineering **626** (2019) 012011 doi:10.1088/1757-899X/626/1/012011

Because the depth of the positioning sleeve is too shallow, which makes the actual size of the cross-section weld very small, and thus reduces the strength of the weld, so under the action of stress, cracks easily occur, and there will be oil leakage in the pressure test station [7].
Because the shallow depth of the positioning sleeve will not only cause the weld strength to be insufficient, but also bring inconvenience to the welders, so the welders must always pay attention to the change of the pool, otherwise, there will be welding deviation and excessive weld height[8].
In the case of welding deviation and excessive weld residual height, there will be welding defects such as non-fusion, which will eventually lead to oil leakage in the weld.
Because the excess height of the weld is too high, the operator will not have the opportunity to repair the weld twice. The final solution is only to go to the lathe and re-weld the weld to solve the problem of oil leakage, which is time-consuming and labor-consuming.

3.2 Inappropriate welding parameters and foot size
For the workers who have been engaged in welding for a long time, most of them are skilled and familiar with the workpiece. Usually, they use 280-300A current and 30-32V voltage to weld, which makes the welding heat input too large and easy to occur such problems as undercut, depression and overheating of heat affected zone. Even in the process of welding, because the welding moving speed is faster than the metal melting speed, the phenomenon that the weld is not fused and penetrated appears. As a result, a large number of welding defects and welding stress concentration will occur, so that the strength of the weld is not enough, and cracks will occur in the weld under the action of stress, resulting in oil leakage.

3.3Inappropriate welding sequence
After controlling the depth and process parameters of the positioning sleeve, the most important thing is to find a correct welding sequence to effectively guarantee the welding quality.
As shown in Fig. 5, the operators usually start the arc at 6 o'clock and complete the welding of the workpiece by three-layer and three-pass welding. Because of the large width of the end weld, there is still a large gap after twice welding at 6 o'clock, but this gap can't be welded twice, so the operator usually only welds once again to complete the welding. As the gap of the last pass is much smaller than that of the first two, it is easy to form a pit when the molten pool metal flows down from the weld in the last pass. At the same time, it is unavoidable that the joint will appear again.

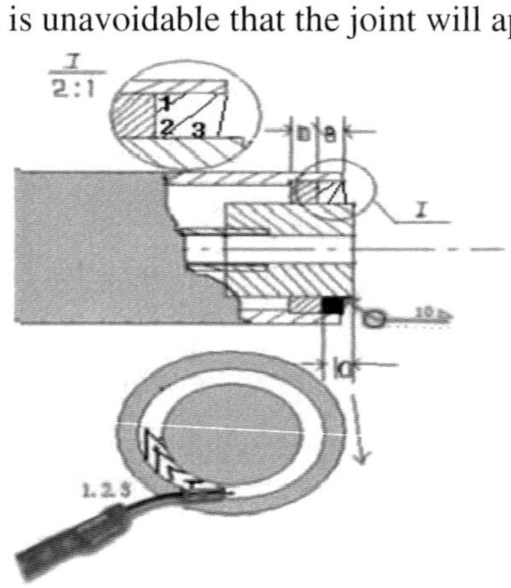

Figure 5. Inappropriate welding sequence

In addition, QY12.16II piston rod should be tested after welding, which belongs to pressure vessel welding. It is taboo to frequently occur joints. Some operators also use four-pass four-layer four-pass welding sequence, but the appearance of the weld after four-pass welding is very poor.
Finally, all operators adopt three-layer three-pass welding method. Although this method has a good shape, compared with four-layer four-pass welding, three-layer three-pass welding is difficult to effectively control oil leakage of piston rod small head rod. Moreover, arc welding at 6 o'clock requires the operator to be steady and observe the molten pool carefully. A slight tremor in the

welding process will be welded to the piston rod, which will lead to the scrap of the piston rod due to de-chromium, so improper welding sequence will lead to oil leakage.

4 Solutions to the Problem

4.1 Strictly control the insertion depth of positioning sleeve

According to the label on the drawing, when the positioning sleeve is placed at a depth of 8 mm, the positioning sleeve does not need spot welding positioning.

Because the matching of positioning sleeve and rod body and rod plug is over-matching, the welding process will not be loose or deformed, so before welding, the depth of positioning sleeve should be strictly checked to ensure that each product positioning sleeve is positioned, and ensure that the positioning sleeve is positioned and not deviated. In this way, the weld seam after welding is also very flat, ensuring that the size and strength of the weld seam specified in the drawings are obtained, preventing cracks and reducing oil leakage.

4.2 Selection of Reasonable Welding Process Parameters

In production, we revised the reasonable welding parameters according to the welding operation instructions and a large number of test instructions, as shown in Table 1.

Table 1 Reasonable welding process parameters

Table 1. Reasonable welding process parameters

	Welding Current) [A]	Voltage [V]	Gas flow rate [L/min]	Number of welding layers	Welding speed [mm/s]
First layer	200-220	22-24	15-20	2	5-6
Second layer	180-200	20-22			4-5

With the above welding parameters, we can ensure penetration, weld size and appearance, avoid undercut and heat affected zone overheating, reduce welding stress and weld crack probability, and improve weld strength and quality.

4.3 Selection of Reasonable Welding Sequence

Since the oil leakage quality accident was discovered, we have carried out theoretical analysis and many welding experiments, and found a good welding method (as shown in Figure 6).

This welding method can effectively reduce the occurrence of oil leakage. Firstly, we change the number of welding layers from three-layer three-pass welding to two-layer four-pass welding. The second layer of welding is beneficial to improve the first layer of weld structure and make up for the welding defects of the first layer. Secondly, we changed the position of the torch to weld 1 weld at 11 o'clock, then 2 weld at 6 o'clock, 3 weld at 11 o'clock and 4 weld at 6 o'clock. The 1 and 3 welds made arc blocking to the rod and increased the fusion of rod plugging and positioning sleeve.The 2 and 4 welds made arc blown to the rod body to promote the rod body and rod body. The fusion of positioning sets. Such a pressure together can ensure compactness and penetration

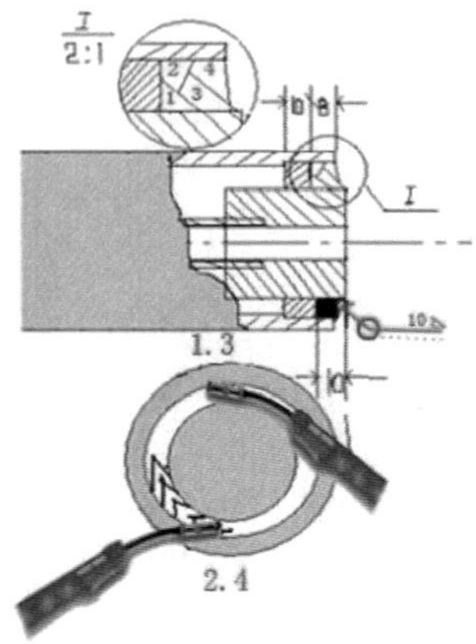

Figure 6. Welding sequence and position after changing

4.4 Preheating before welding and heat preservation after welding

In order to reduce the cracks of 45# steel during welding, workers are urged to take pre-heating and post-welding insulation measures strictly according to the process requirements .The specific approach is to use oxygen acetylene flame to preheat 100~150℃, then welding. Local heater is used for 0.5-1 hour after welding, and the holding temperature is 250-280 ℃. The cooling rate is delayed, the hardening degree of weld and heat-affected zone is reduced, the crack generation is reduced, and the oil leakage caused by crack is greatly reduced.

5 Effect of measures taken

After taking the above measures, the quality inspectors randomly sampled 60 pieces of small head rod plug of QY12.16II piston rod. The test results are shown in Table 2. From Table 2, we can see that the oil leakage rate of QY12.16II type decreased from 20% to 5% after taking measures, which effectively controlled the oil leakage of plugging weld of piston rod small head rod. It shows that the above measures are feasible.

Table 2 Sampling results before and after taking measures

	Total production	Number of oil spill products	Oil leakage ratio
Before improvement	60	12	20%
After improvement	60	3	5%

6. Conclusion

In this paper, we mainly study the quality control measures of plugging weld of small end rod. After analyzing the reasons that affect the weld quality, such as shallow insertion depth of piston rod positioning sleeve, welding sequence, welding process parameters and small foot size, we take measures to change the welding sequence and position, strictly control welding process parameters, improve the quality consciousness of welding operators, and control the depth of positioning sleeve. Through the above measures, the quality of our welds has been greatly improved, and the problem of oil leakage has also been solved.

Acknowledgments

Thank you for all the scholars involved in this paper. This paper quotes the research literature of several scholars. Without the help and Inspiration of the research results of various scholars, I would be very difficult to complete the writing of this paper.

Thank my colleagues and family members for giving me a lot of material in the process of writing my paper, and also for providing enthusiastic help in the process of writing and typesetting.
Because of my limited academic level, there are inevitably some deficiencies in the papers I have written. I urge teachers and editors to criticize and correct them.

Reference

[1] Jing Xuwen, Xu Zhenzhong, Li Huan， Performance optimization of diesel engine piston based on multidisciplinary coupling [J].*Mechanical design and manufacturing.* 2009 (06)
[2] Wang Jianping, Hu Dingyun, Wu Bo， Fatigue Life Prediction of Diesel Engine Aluminum Alloy Piston [J]. *Mechanical Design and Manufacture.* 2009 (12)
[3] Wang Xianfen, Liu Xiangfa, Ding Haimin， Phase analysis of high performance Al-Si-Cu-Ni-Mg piston alloy [J]. *Casting.* 2008 (02)
[4] Li Peng. Present situation and development trend of piston design for automobile engine [J]. *Automotive technology and materials,* 2008 (01)
[5] Liang Mu, Yang Baoguo， Discussion on finishing technology of piston inner surface [J]. *Journal of Zhongyuan Institute of Technology.* 2008 (03)
[6] Hu Huifang, Li Huaji, Xue Hansong, Kongfanshi， High temperature properties of high silicon piston alloys based on orthogonal regression [J]. *Casting technology.* 2008 (08)
[7] Ling Haomin, Fan Jianxin， Rational design of oscillating cooling structure for combined piston [J]. *Diesel engine.* 2008 (04)
[8] Liang Gang, Chen Zhizhong, Yang Zhixiang， New Development of Piston Design Technology for Medium Speed and High Power Diesel Engine [J]. *Diesel Engine.* 2007 (02)

IOP Conf. Series: Materials Science and Engineering **626** (2019) 012012 doi:10.1088/1757-899X/626/1/012012

Talking about New Technology and Equipment of Automatic Welding Application of Construction Machinery in the Future

Min Hu

Jiangsu province Xuzhou technician institute, Xuzhou, Jiangsu 221151, China

331947080@qq.com

Abstract: This paper introduces the working principle and application of new automatic welding technology and welding equipment, analyzes the possibility of its application in construction machinery field, and expects to provide beneficial reference for improving the welding quality of XCMG machinery, reducing production cost and strengthening brand effect.

1. Introduction

The global manufacturing industry continues to shrink, the market has a lot more than the market demand, the construction machinery industry is facing an unprecedented winter. Taking a broad view of the market, the whole equipment manufacturing industry has entered into a state of dismal operation, the world's major well-known construction machinery industry scale and the corresponding measures are foreshadowing the future development of the industry will still be in the winter period. In the face of such a severe market situation, improving product quality, reducing production and use costs, and widening the gap in brand effect are the keys to winning the market tug of war. After more than two centuries of continuous development in the construction machinery industry, the threshold of traditional manufacturing technology has been greatly lowered, and production automation, lightweight products and energy-saving use have gradually become the focus of future competition [1]. As a leading enterprise of construction machinery in China, Xu gong group is deeply aware of the crisis of future development. It takes the lead in introducing industrial robots in batch and is widely used in cutting, welding, painting, assembly and other fields. The product quality and production efficiency are significantly improved.

The introduction of industrial robots will undoubtedly improve the new heights of engineering machinery manufacturing, industrial robots also has its drawbacks, however, in the case of welding robot, the material is expensive, lamella piece can't accurate arc tracking, easy to deformation structure cannot be accurately using multilayer multichannel welding, thin wall tube joint is no fusion welding, narrow groove root is poorer.

2. Laser tracking technology

The key of robot automatic welding is intelligent position finding and arc tracking technology. At present, most of the intelligent locating technologies of welding robots mostly use welding wire contact locating, and some welding robots choose nozzle contact locating, such as IGM welding robot [2].

The robot's contact finding circuit is a normally open circuit. When the welding wire or nozzle touches the work piece, the normally open circuit closes and the robot's built-in relay absorbs and

IOP Conf. Series: Materials Science and Engineering **626** (2019) 012012 doi:10.1088/1757-899X/626/1/012012

closes. The processor records the position information of this contact point. In the process of locating, the deviation between the actual position of the reference point on the work piece surface and the programming position can be determined, and the corresponding welding procedure can be corrected. When the welding contact finding error occurs, the position of welding seam will also change with the robot's automatic correction, which is the most fundamental reason for the robot's automatic welding to produce welding defects. Take the position finding of welding wire as an example. In order to ensure the position finding accuracy, it is necessary to control the dry elongation of the position finding of welding wire. With the aging of wire clip gun cable and the blocking of wire clip cylinder, the reliability of wire clip decreases gradually. In the face of thin plate below 3mm or the working condition of front finding and back welding, the change of dry elongation can easily lead to the finding deviation, and then the welding defects appear. In addition, there are many factors affecting the positioning accuracy of the wire contact, such as the wear of the conductive nozzle, the surface quality of the work piece, the wire conductivity and so on. Similarly, the contact location of the nozzle has its drawbacks. The nozzle is prone to thermal deformation and circularity loss, and the nozzle is prone to adhesion and welding spatter, which reduces its conductivity and so on.

At present, the seam tracking technology used in automatic welding of Xu gong Group is arc-tracking technology. When welding thick plate or fillet weld, the torch swings horizontally. The difference of dry elongation leads to the difference between the actual welding current and the set current. The shorter the dry elongation, the larger the actual current, and the longer the dry elongation, the smaller the actual current. Using this principle, the corresponding software processes the detected current change and the position of the welding gun in real time, so as to correct the actual trajectory of the robot, ensure that the center line of the trajectory is always in the middle of the groove, and ensure that the height of the welding gun is consistent with that of the welding seam. Among the various droplet transition forms, jet transition welding has the advantages of stable process, less spatter, large depth and beautiful weld formation, which is the ideal arc tracking state. In order to ensure a good arc tracking state, the actual welding current is greater than the minimum current for the transition from droplet to jet. Taking the welding wire of 1.2mm ER50-6 and mixed gas shielded arc welding as examples, the critical current is 240A. When the welding current is lower than 240A, the arc tracking characteristics tend to be unstable. Therefore, arc-tracking technology is suitable for welding of high current and medium thickness plate structure [3]. When the thickness of plate is less than 5 mm, the arc tracking accuracy begins to decline, and when the thickness of plate continues to fall below 3 mm, the arc tracking function fails. Another important factor that makes thin sheet metal unsuitable for arc tracing technology is that the edge with smaller lap edge is prone to overheat during welding. The higher the temperature, the larger the resistance and the smaller the welding current. When the arc tracking is used, the torch will automatically correct in the direction of the smaller current, which will lead to the deviation of the bead and the melting of the base metal with short lap edge. In addition, the surface quality of weld, joint quality, chemical composition of welding wire and mixed gas will affect the arc tracking effect.

The application of laser tracking technology can effectively solve the above problems. Laser tracking is a technology that laser irradiates a moving target and controls the direction of the measurement system according to the deviation angle between the laser signal reflected from the target and the optical axis of the measurement system. Laser tracking system usually consists of laser illuminator, laser detector, information processing system and servo system. The laser signal emitted by the illuminator to the target is reflected back by the target and projected onto the detector through the optical system. The detectors are generally four quadrant photodetectors. If there is a deviation angle between the reflected signal of the target and the optical axis of the measuring system, the area of the spot projected on the photoelectric detector is not equal in four quadrants. After the photoelectric signal is processed, the corresponding angular error signal is obtained. Using this signal to drive the servo system, the measurement system can turn to the target until the deviation angle is zero, thus realizing laser tracking. The laser tracker developed at present can not only realize efficient seam tracking, but also realize intelligent seam location. Laser positioning can not only make use of laser ranging technology to contact seam positioning, but also use laser-imaging technology to measure the height and transverse position deviation of the seam at one time. Five can reduce the non-working time of the robot-eight times.

Laser position finding and laser tracking technology can solve the problem of variation of dry elongation and fluctuation of welding current, but it requires higher quality of work piece surface and joint precision [3]. Rough weld surface or fluctuating joint gap will generate diffuse reflection in laser imaging, thus reducing the laser tracking precision. Because of the relatively large volume and high cost of the laser tracker. It can be generally used as a supplement to the contact position finding and arc tracking technology, and applied to the welding of large open structural parts or thin plates.

3. Welding deformation control technology

During welding, the molten metal and base metal in the near-seam zone are heated and expanded, resulting in plastic deformation. During solidification, the metal in the weld and near the seam shrinks, resulting in longitudinal and transverse internal stress, longitudinal and transverse contraction of the weld and deformation of the weldment. Since heat expansion and cold contraction are inevitable during welding, welding deformation is inevitable. We can only control but cannot eliminate welding deformation.

There are many factors affecting the welding deformation, such as the nature of the base metal, weld form, welding heat input, welding direction, joint gap, and amount of molten metal and so on. Under certain welding environment, reducing welding heat input and optimizing welding sequence is undoubtedly the most effective way to control welding deformation. The following introduction of welding technology provides a new direction for reducing welding deformation.

Cold arc welding technology: With the advance of industrial process, thin plate and light metal structure, parts are more and more widely used [4], ultra-thin galvanized plate has been used in batch welding of cab and fuel tank. In order to reduce plate-welding deformation and avoid plate-overheating breakdown, the welding heat input of this kind of plate must be controlled.

Short-circuit transition welding with lower current and voltage is the best choice for traditional thin plate welding. The droplet contacts with the molten pool directly before leaving the end of the wire. The arc instantaneously extinguishes the short circuit. The droplet transits into the molten pool under the action of electromagnetic contraction force produced by the short circuit current and surface tension of the liquid metal. Now of short circuit, the voltage drops sharply and the current rises gradually as the resistance decreases.

Now of dropping, the voltage rises sharply to ensure the resignation of the arc, while the current can only fall slowly under the action of inductance. At this time, the current and voltage are high, and welding defects such as arc instability, undercut, welding spatter and breakdown are easy to occur. Cold arc welding technology is to make use of digital signal processor to make quick adjustment according to the feedback current and voltage now of arc returning, to ensure the stability of arc returning.

After arc initiation, a small pulse current is given to make the arc obtain higher arc energy and promote the re-nucleation of melt droplets. The cold arc welding technology makes the melting droplet transition period stable, guarantees low heat input, and greatly reduces the welding spatter caused by traditional short-circuit transition.

Double wire and triple arc welding technology: the welding parts in the construction machinery industry are mainly medium and thick plates, and double wire automatic welding can effectively improve the welding efficiency. It is widely used in the welding of key welding seams such as loading mobile arm, big arm of excavator and crane [4].

While increasing the welding efficiency, the welding heat input is also greatly increased. Thermal deformation after welding has always been a difficult problem in the construction machinery industry. Different from traditional twin wire welding, twin wire three arc welding can form a third arc between twin wire arcs [5]. In the twin-wire three-arc welding system, the arc power used to melt the wire is much larger than that used to melt the base metal. The melting efficiency is greatly improved, and the welding heat input is significantly reduced. The arc power between three arcs can be allocated by intelligent control of digital signals. When high melting depth is needed, the arc power for melting base metal can be increased, the arc power for melting wire can reduced [6], and enough heat input can be guaranteed for melting base metal. Similarly, when it is necessary to control welding deformation and improve welding efficiency, the arc power used to melt base metal can reduced, and the arc power used to melt welding wire can increased to ensure sufficient melting efficiency.

Fish-scale welding technology: the difficult problem of thin-wall pipe welding lies in the control of welding attitude, especially the welding of through-line pipe wall joint. The gas shielded welding in the form of gag; brazing or short circuit transition is widely used in the traditional welding of this kind of welding seam. It is difficult to guarantee the welding strength of brazing due to the high cost of tag welding [7], it is very difficult to control the heat input and output of welding during vertical welding and horizontal welding of gas shield welding, and it is very easy to break through the pipe wall. The introduction of fish scale welding technology provides a possibility for automatic welding of bridge handrail and tower crane support frame. During the whole process of fish scale welding, the robot always makes a step movement. For each distance the robot advances, it will stay in the same place for a period, then advances for a period, and then stays for a period, and so on until the end of welding. When the robot stops, the welding power supply will start arc welding quickly [8], and when the robot moves, the welding machine will shut down instantly. The welding void ratio of fish-scale welding per second is 35%~55% (generally 40%), which greatly reduces the overall welding heat input compared with the traditional welding process. It can realize all-position welding, and the welding seam presents circular flake fish-scale pattern. At present, it has been widely used in the welding of bicycle and motorcycle frames, fitness equipment and medical equipment.

4. Summary

The welding equipment and welding technology described in this paper are only the tip of the iceberg of many new welding technologies. Welding automation technology is a comprehensive subject integrating welding, automation, machinery, hydraulic pressure and measurement, which requires stable pre-welding preparation, skilled process monitoring and professional performance maintenance. Then the advanced welding technology, if there is no suitable structure, mature welding process, stable quality, excellent quality of personnel can not get the ideal welding quality. In today's increasingly fierce competition, the traditional essence must be precipitated internally, and the advanced technology must be absorbed externally, to help XCMG to improve the new manufacturing level.

Reference

[1] Zhang bo. Application analysis of automatic welding equipment in construction machinery manufacturing [J]. *Internal combustion engine and accessories*, 2018(14).

[2] Tian peng, liao qiuqin. Research on application of automatic welding equipment in construction machinery manufacturing [J]. *China high-tech enterprise*, 2016(14):45-46.

[3] Zhang xiaoyu, zuo jian, yu ruibin. Discussion on welding automation technology of construction machinery [J]. *Science and technology prospect*, 2016, 26(14).

[4] Song hengchang, liu feifei. Application of welding and cutting equipment in efficient welding of construction machinery manufacturing [J]. *Chemical industry management*, 2018, No.487(16):160.

[5] Meng Qingdi, Wang Xinge, Han Huiwei. Application and Development of Laser Technology in Aviation Equipment Support[J]. *Mechanical Design & Manufacturing*, 2009(6): 258-260.

[6] Anonymous. Into the welding robot application site [J]. *Metal Processing (Hot Working)*, 2014 (10): 24-27.

[7] Chen Hengfeng, Sheng Hui, Guo Hui, et al. Development of Industrial Welding Robots[J]. *Agricultural Science & Technology*, 2016(5): 60-61.

[8] Du Bing, Guo Wei. Development direction of high-end equipment manufacturing and welding technology [J]. *Welding*, 2017.

IOP Conf. Series: Materials Science and Engineering **626** (2019) 012013 doi:10.1088/1757-899X/626/1/012013

Development of a New Fluid Shearing System for Cell Culture in Vitro

Pengfei Wu, Xin Wang, Lilan Gao and Chunqiu Zhang*
Tianjin Key Laboratory for Advanced Mechatronic System Design and Intelligent Control, Experimental Mechanical and Electrical Engineering Education, Tianjin University of Technology, Tianjin, 300384, China

13012251281@126.com

Abstract. Cells in organisms are surrounded by fluid environment, and the fluid shear force produced by this fluid affects the morphology and behaviour of cells. The stimulation of cells in a bioreactor is a key area of research in tissue engineering and regenerative medicine. This paper introduces a fluid shear system for cell culture in vitro. The main component of system is a perfusion chamber, which consists of an inoculation room and a perfusion cap. The inoculation room is used for the common laboratory consumables of microorganism or cell culture, such as petri dish, six-well plate and so on. It can realize to six independent experiments of fluid shear; the new type of perfusion cap was designed to obtain a reasonable cavity by hydrodynamic simulation. The cap was produced by casting technology. It not only has the function of cap with nozzles, but also has the function of sealing. With peristaltic pumps, nozzles, liquid storage bottles and so on, the fluid shearing system which can provide uniform, controllable and up to 60 dyn/cm^2 shear stress along the bottom surface is constructed, and is easy to operate and can be personalized.

1. Introduction

In normal case, cells are cultured in a relatively static state in the laboratory with controlling environmental variables such as CO_2 and PH to simulate cell culture in vivo. But In the real case, many cells are exposed to shear stress caused by the flow of a bio-fluid system. For example, vascular endothelial cells in blood vessels are subjected to shear of blood flow and hydrostatic pressure. These mechanical forces have a great influence on the physiological response and adhesion of cells [1]. There is a fluid shear force loading device for studying cells in vitro in laboratory. Planting cells on culture plates, the culture fluid is pumped by a power source and flows through the culture plate to stimulate cell growth [2]. Its core consists of three parts: an experimental part consisting of two parallel plates and rubber gaskets clamped in the middle for sealing; a perfusion system consisting of a peristaltic pump, a reservoir and a conduit providing a power source and an observation system for observing cell growth. Then, Andrew et al. [3] developed a parallel plate flow chamber. The flow chamber has two compartments: one compartment contains cells, and the other contains NO sensitive motors. The relationship between NO produced by endothelial cells and vascular tension was also studied. Afterwards, Wong et al [4] introduced a parallel plate-flow chamber to describe the mechanical properties of endothelial cells exposed to laminar shear stress. Based on a lumped parameter hemodynamic model, Wang et al [5] designed and fabricated a parallel-plate flow chamber system with appropriate components. Beyond that, Utsunomiya et al [6] investigated the effects of fluid shear stress on cultured human corneal epithelial cells with the parallel-plate type of flow chamber, they found Wound healing, cellular proliferation and growth factor expression. Jahn et al [7] used the flow chamber model to detect the antiplatelet effect of clopidogrel. A microfluidic approach have also been reported [8] with assess mechanically regulated osteocyte-osteoclast communication. In recent years, researchers have been studying the effects of shear fluids such as inlet and outlet optimization, and unsteady flow on cell growth on the basis of parallel plate flow chambers.

IOP Conf. Series: Materials Science and Engineering **626** (2019) 012013 doi:10.1088/1757-899X/626/1/012013

Previous perfusion system is an important experimental device and means to study cells in vitro with parallel plate flow chamber, which is not universal for the size, and is designed according to the needs of its own experiments. Upper and lower flat plates need rubber ring seals, more assembly parts, cumbersome operation, high accuracy requirements. This paper introduces a new and convenient perfusion system. Based on common laboratory consumables, such as Petri dish, six orifice so on, a new type of irrigation cap matched with Petri dish has been developed. With six orifices, up to six independent parallel experiments can be realized. It has the characteristics of simple operation and personalized customization. Establishment of a fluid shear stress system for cultured cells in vitro requires connect peristaltic pump, nozzle, liquid storage bottle, joint, etc. It is of far significance for culturing cells in laboratory and exploring the biomechanical response of cells.

2. Models and methods

The standard Petri dish with a diameter of 35 mm and a height of 10 mm is used in the inoculation room of the system. The flow chamber structure diagram is shown in Fig.1. Assuming that the fluid is incompressible Newtonian fluid, its density is $p=1000 kg/m^3$, and the flow state is laminar. The Gambit software establishes geometric model, generates structured hexahedron mesh and sets the inlet speed as a boundary condition. Secondly, import the model into FLUENT of CFD software, the relationship between the inlet velocity and the undercut stress was studied with changing feed speed. The specific values of the geometric dimensions of the flow chamber used for calculation are listed in Table 1.

Figure 1. Systematic flow chamber structure diagram

Table 1. Geometric dimensions of flow chambers used for calculation

L (mm)	L1(mm)	L2(mm)	L3(mm)	W(mm)	H(mm)	h(mm)	R(mm)	D(mm)
35	24	2.6	10	12	1	0.3	17.5	2.8

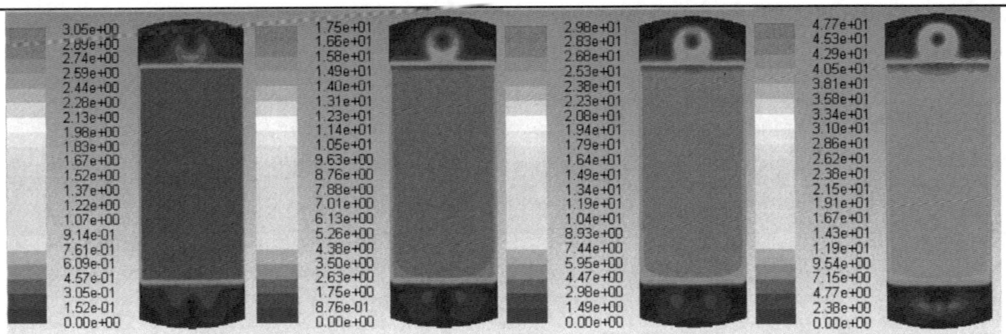

IOP Conf. Series: Materials Science and Engineering **626** (2019) 012013 doi:10.1088/1757-899X/626/1/012013

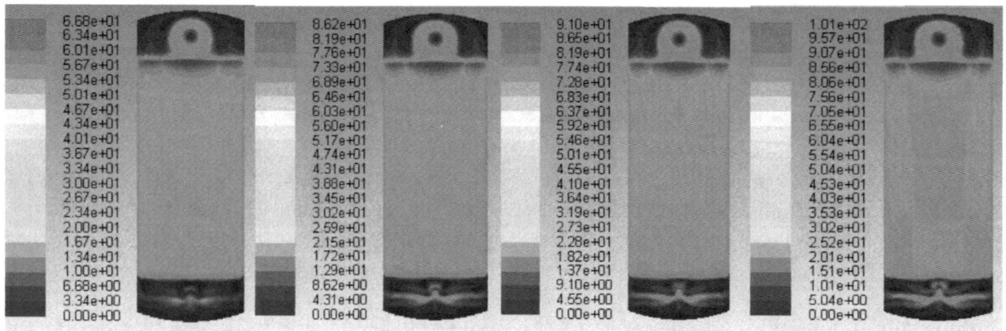

Figure 2. Bottom shear stress nephogram with entrance velocity of 0.01m/s, 0.05m/s, 0.08m/s, 0.12m/s, 0.16m/s, 0.20m/s, 0.21m/s and 0.23m/s

Fig. 2 shows the bottom shear stress nephogram at different inlet velocities. The shear stress distribution in the middle of the bottom surface is uniform when the inlet velocity is between 0.01m/s and 0.2m/s, the uniform range of shear force in the middle of the bottom surface is gradually reduced when the inlet flow rate is greater than 0.2 m/s.

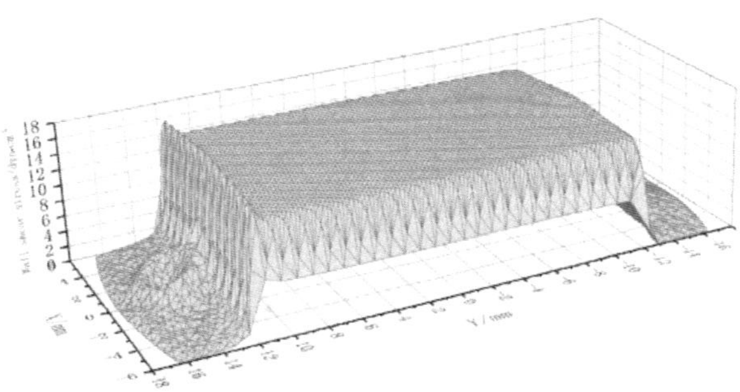

Figure 3. Shear stress distribution on the bottom when the inlet velocity is 0.05 m/s

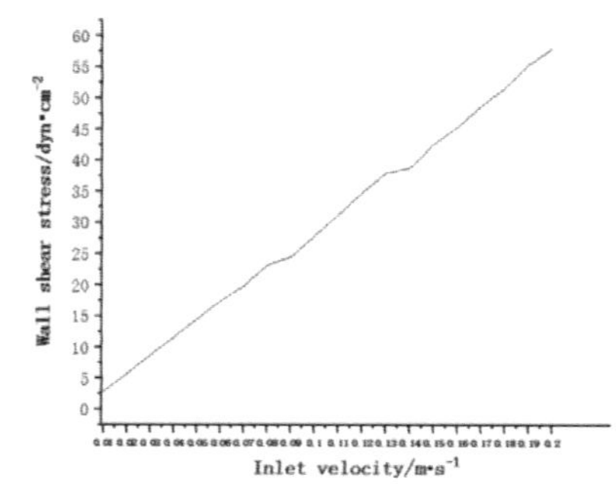

Figure 4. Relationship between inlet velocity and bottom shear stress

The shear stress distribution at the bottom of the flow chamber is shown in Fig 3 when the inlet velocity is 0.05m/s. It can be seen that shear stress changes take place near the entrance and exit as well as the side wall, Most of the areas in the middle are uniform. Fig. 4 shows the relationship between velocity and bottom shear stress when inlet velocity is 0.01m/s-0.2m/s. This provides a theoretical basis for subsequent experiments.

3. System building

The sketch of the system designed in this paper is shown in Fig.5. The main components of the irrigation system are as follows: Peristaltic pump (TL-LL-100W), perfusion chamber, conduit (19 #), multi-way tube joint (3-way), liquid storage bottle (200 ml), etc. The Peristaltic pump provides stable power source for liquid flow, and the perfusion chamber consists of the inoculation room and the perfusion cover. The inoculation room is constructed by common consumables Petri dishes and six-hole plates in the biological laboratory. The perfusion cover is made of addition-moulded silica gel. Because of its complex shape, it is prepared by casting moulding technology, and multi-way tube joint can realize multi-channel parallel flow. The shape of the designed die is square and the inner cavity is cylindrical, the bottom of the chamber is the designed flow field slope. The center of the mold cover is provided with a spout and four exhaust ports around it, the advantage of this design is that the air left in the chamber is discharged from the periphery during pouring. The lower side of the cover has a convex platform, and the upper side of the base has a corresponding concave platform, it is not only plays a positioning role, but also plays a pressing role. Molds are made by 3D printing technology. Additional silica gel A and B with mass ratio of 1:1 are mixed and stirred, and then put into vacuum injection moulding machine for air extraction, when the air is removed, it is placed in a drying thermostat for vulcanization. The design of the mold, the printed body and

IOP Conf. Series: Materials Science and Engineering **626** (2019) 012013 doi:10.1088/1757-899X/626/1/012013

the formed perfusion cover are shown in Figure 6. Seeding the cells in the vaccination room, connecting peristaltic pump, liquid storage bottle, conduit, multi-way tube joint and perfusion chamber in turn, opening peristaltic pump and adjust parameters, the perfusion fluid is pumped out by a peristaltic pump and then flowed into the flow chamber through a pipeline which applying shear stress to cells. Then it is piped back into the liquid storage bottle to form a circulating system. The fluid shear force system for culturing cells in vitro is constructed as shown in Fig. 7.

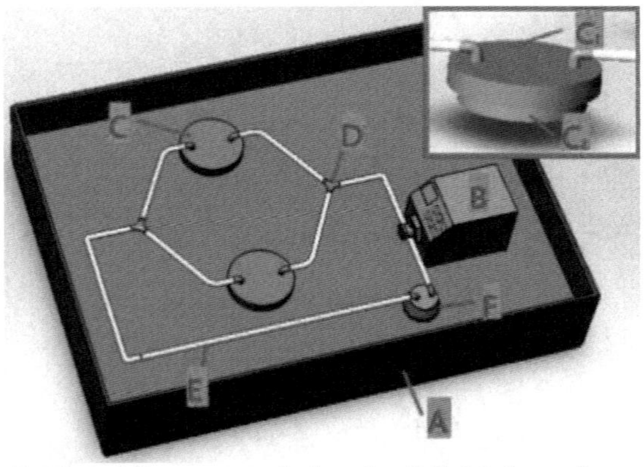

Figure 5. Three-dimensional sketch of fluid shear force system
A: Workbench, B: Peristaltic pump, C: Perfusion chamber, C_1: Perfusion cover, C_2: Inoculation room, D: three-way tube joint, E: Catheter, F: Liquid storage bottle.

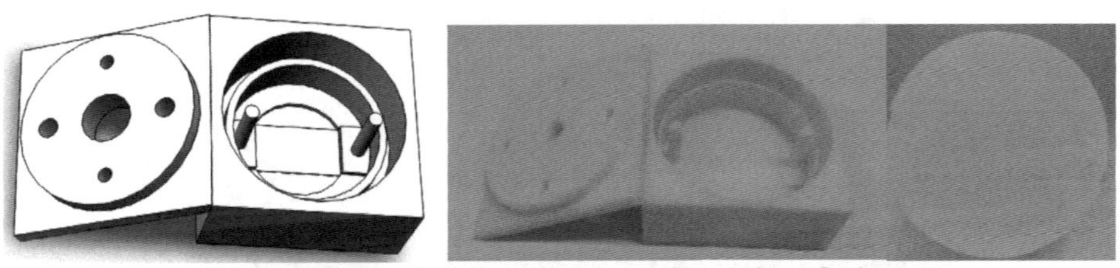

Figure 6. Die design, printing entity and perfusion cover

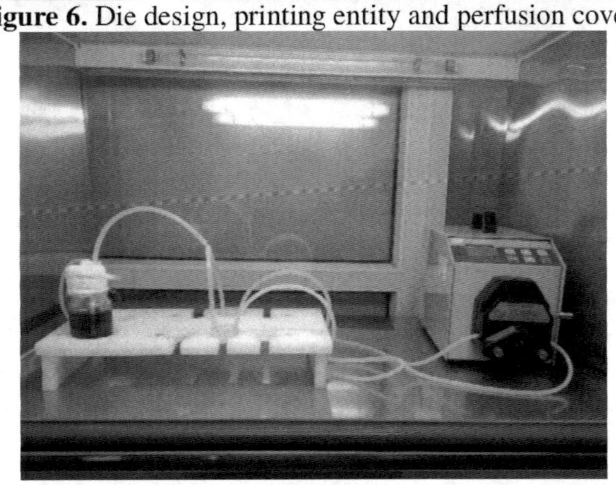

Figure 7. The fluid shear force system for culturing cells in vitro

4. Conclusion

The fluid shear force system of cells in vitro was constructed for the study of cell mechanics and biology in vitro. The main conclusions are as follows: the novelty of the system lies in the formation of perfusion system based on general laboratory consumables. The inoculation chamber is made of common laboratory consumables, such as Petri dishes or six-hole culture plates, which can give full play to the convenient role of cell culture in this system. The perfusion cap is made by pouring silicon collagen solution, which has standard size and can be directly used with Petri dishes. The system can independently study six cell samples at the same time; it has strong versatility and simple operation. Through hydrodynamic analysis

IOP Conf. Series: Materials Science and Engineering **626** (2019) 012013 doi:10.1088/1757-899X/626/1/012013

and the feasibility of the detection device, the theoretical uniform shear stress can be obtained and a more reliable fluid shear stress system for cultured cells in vitro can be provided.

Acknowledgments

The project was partly supported by the National Natural Science Foundation of China (11672208, 11432016, 31700812 and 11572222,) and Tianjin Natural Science Foundation (18JCZDJC36100, 18JCYBJC95200, and 18ZXSGSY00010).

Reference

[1] Chiu JJ, Chen CN, Lee PL, Yang CT, Chuang HS, Chien S and Usam S 2003 *Journal of Biomechanics*. **36** 1883-1895.

[2] Frangas JA, Eskin SG and Mcintire LV and Ives CL 1985 *Science*. **227** 1477-1479.

[3] Andrews AM, Jaron D, Buerk DG, Kirby PL and Barbee KA 2010 *Nitric Oxide-Biology and Chemistry*. **23** 335-342.

[4] Wong AK, Llanos P, Boroda N, Rosenberg SR and Rabbany SY 2016 *Cellular and Molecular Bioengineering*. **9** 127-138.

[5] Wang YX, Xiang C, Liu B, Zhu Y, Luan Y, Liu ST and Qin KR 2016 *Biomedical Engineering Online*. **15** 659-672.

[6] Utsunomiya T, Ishibazawa A, Nagaoka T, Hanada K, Yokota H, Ishii N, and Yoshida A 2016 *Investigative Ophthalmology & Visual Science*. **37** 6382-6388.

[7] Jahn K, Suchodolski K, Schafer A, Sahlmann B, Kuster U, Echtermeyer F, Calmer S, Theilmeier G and Johanning K 2017 *Anesthesia Analgesia*. **124** 1091-1098.

[8] Middleton K, Al-Dujaili S, Mei X, Gunther A and You L 2017 *Jouranl of Biomechanics*. **59** 35-42.

IOP Conf. Series: Materials Science and Engineering **626** (2019) 012014 doi:10.1088/1757-899X/626/1/012014

Experimental Study of Acoustically Enhanced Heat-Transfer from Copper Sphere in a Steady Flow Field

Xu Weilong[1,*], Jiang Genshan[2], Peng Weicai[1], He Xuesong[1]

[1]National Key Laboratory on Ship Vibration & Noise, China Ship Develop and Design Center, Wuhan 430064, China

[2]Department of Mathematics and Physics, North China Electric Power University, Baoding 071003, China

milo0324@163.com

Abstract. The combined effects of acoustic field and flow field on the forced convective heat transfer rate during cooling of a stationary sphere were investigated experimentally. The acoustic frequency is in the range of 100 to 2500 Hz, with the velocity ratio of acoustic velocity and flow velocity is in the range of 0.59 to 4.76. The relationship among sound pressure levels, frequency, flow velocity and the diameter of the sphere with the temperature gradient of the sphere were analyzed. The results show that higher velocity ratio cause higher heat-transfer rate. However, when the sphere size is 5 mm, the heat transfer rate is always lower than that without an acoustic field during our experiment, and when the sphere size is 10 mm and the velocity ratio is higher than 3.38, the heat transfer rate is higher than that without an acoustic field, but when the size of the copper sphere is 15 mm, the velocity ratio is decreased to 1.89. There is a special frequency, at which *Nu* reaches a maximum. This special frequency increases with increasing SPLs. The conclusions highlight a good application prospect in combustion for the boiler.

1. Introduction

In the power plant boiler, coal particles are carried by the flue gas and injected into the furnace for combustion, the particles are either stationary or state of motion with a certain slip velocity with respect to the flue gas. In recent years, many studies have shown that the rates for heat and mass transfer from an object in an oscillating fluid can be enhanced significantly [1].

The effect that an oscillating flow-field has on the heat- and mass-transfer from single spherical particles and droplets has been investigated since the late 1930s[2]. Because the oscillating flow is caused using mechanical methods, the frequency-range in these studies, however, is very narrow. An acoustically driven oscillation flow (because it is a vibration flow) can produce not only a larger amplitude but also a wider frequency-band. Larsen and Jensen[3] investigated the evaporation rate of a single droplet of distilled water within a sound field. The Reynold number based on the oscillation velocity varied between 16 and 180, with sound pressure levels from 132 dB to 152 dB and frequencies between 82 and 734 Hz. Ha and Yavuzkurt[4] studied theoretically the acoustic enhancement of heat- and mass-transfer from particles and droplets with a size of 100 μm. The velocity and temperature fields around the particle for different oscillating flows were analyzed in detail. The Reynold number varied between 15.7 and 94.4, with sound pressure levels from 151 dB to 167 dB, and frequencies of 50 Hz, 1 kHz and 2 kHz. Gopinath and Mills[5] investigated the flow field and convective heat-transfer rates from a sphere due to acoustic streaming from an isolated sphere in a standing sound field. Analytical and numerical solution techniques were used to obtain Nusselt number correlations for a wide range of Prandtl numbers. Alassar[6] investigated both forced and mixed convection heat transfer from a sphere with an axisymmetric oscillating flow. The Navier-Stokes equations for incompressible fluids were solved using a series truncation method, where the stream-function, temperature, and the vorticity were expanded in terms of Legendre and first associated Legendre functions. Hossein and Da-Wen Sun[7, 8] studied (experimentally) the transfer, in particular the heat exchange at the surface, under the effect of ultrasound irradiation around a stationary copper sphere during cooling. The frequency of the ultrasound was 25 kHz. Visual observations revealed that both cavitation and acoustic streaming are important for heat transfer. Xu and Jiang[9] studied the acoustically enhanced heat transfer from a single particle in flue gas in both numerical and experimental methods. The numerical and experimental investigations indicated that the acoustic field can increase higher heat-transfer significantly. Higher sound pressure levels cause higher heat-transfer rates.

The aforementioned studies indicate that higher sound intensities increase the rate of heat and mass transfer significantly. There are no corresponding conclusions, with respect to the combined effect of acoustic field and flow field, however. Therefore, the aim of this study was to investigate the combine effect of acoustic fields and flow fields on heat transfer between coal particles, which move with a certain slip velocity relative to the flue gas. The copper spheres were used to evaluate the convective heat-transfer studies. A lumped system is likely to be the result when copper is used in these experiments. The effect of thermal conductivity becomes negligible thanks to the use of copper spheres in the experiments.

2. Heat-transfer equation

The heat-transfer equation for one-dimensional heat-transfer using spherical coordinates for a solid sphere in air can be expressed as follows:

$$\frac{1}{r^2}\frac{\partial}{\partial r}\left(r^2\frac{\partial T}{\partial r}\right)=\frac{1}{\alpha}\frac{\partial T}{\partial t} \tag{1}$$

where r is the distance from the center of the sphere along the radius, α is the thermal diffusivity, T is the temperature, and t is time.

To study the temperature gradient within an object, the Bi number is calculated as follows:

$$Bi=\frac{hD}{\lambda} \tag{2}$$

where h is the convective heat-transfer coefficient (W • m^{-2} • K^{-1}), D is characteristic length (m) and λ is the thermal conductivity (W • m^{-1} • K^{-1}). If the Bi number is lower than 0.1, the temperature gradient across the object can be neglected.

The initial and boundary conditions for Equation (1) can be described as follows:

Considering the initial condition (t=0):

$$T(r,0)=T_i \tag{3}$$

The boundary conditionscondition, for t>0 is:

$$\begin{cases}\dfrac{\partial T(0,t)}{\partial r}=0\\ -\lambda\dfrac{\partial T(R,t)}{\partial r}=h[T(R,t)-T_\infty]-Q_S\end{cases} \tag{4}$$

Assuming a low Bi number:

$$\frac{\partial T}{\partial r}=0 \tag{5}$$

Q_S in Equation (4) is the amount of heat transferred into the sphere via acoustic radiation. It was assumed that after radiation of the acoustic field, while the sphere is in thermal equilibrium at room temperature, the heat generated at the surface of the sphere cause the temperature of the sphere to increase.

Based on the above conditions, there is a convective boundary at the surface of the sphere, which is proportional to h and the differencedifferences between the surface and air temperature. In addition, heat generation may occur at the sphere surface. The analytical solution of Equation (1) can be expressed as[3]:

$$T = \frac{\left(\exp(\frac{-hA}{\rho V C_{ps}} t)[h(T_i - T_\infty) - Q_S]\right) + hT_\infty + Q_S}{h} \tag{6}$$

or

$$\frac{T - T_\infty}{T_i - T_\infty} = \frac{\left(\exp(\frac{-hA}{\rho V C_{ps}} t)[h(T_i - T_\infty) - Q_S]\right) + Q_S}{h(T_i - T_\infty)} \tag{7}$$

where T is the temperature (°C), T_∞ is the temperature of the air (°C), T_i is the initial temperature of the sphere (°C), c_{ps} is the specific heat of the sphere ($J \cdot kg^{-1} \cdot K^{-1}$)and A is the surface of the sphere (m^2), and V is the volume of the sphere (m^3). Plotting T vs t, according to Equation (6), can be utilized to calculate h. The curve-fitting tool in MATLAB was used to fit the experimental data using Equation (6). Then, the dimensionless number Nu can be expressed as:

$$Nu = \frac{hD}{\lambda_f} \tag{8}$$

where λ_f is the thermal conductivity of the cooling medium, which was 0.0267 $W \cdot m^{-1} \cdot K^{-1}$ for air. The heat-transfer enhancement factor can be calculated using:

$$\text{Enhancement factor} = \frac{Nu_S}{Nu_0} \tag{9}$$

where Nu_s is the Nusselt number calculated when both the acoustic field and flow field were applied and Nu_0 is the Nusselt number without an acoustic field.

3. Experiment and Result Analysis

3.1. Experiment set-up

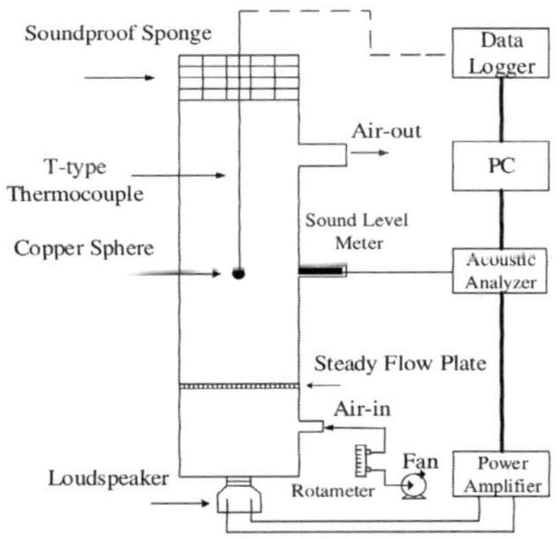

Figure 1. Schematic drawing of the experimental setup

Figure 1 shows a schematic of the experimental setup. A 500 W, 8Ohm loudspeaker is used to produce the sound field. It is positioned under the waveguide tube. The frequency and intensity are controlled using an Acoustic Analyzer (MC3642, BSWA, China), and amplified with an FdB-Model TA1100 professional power amplifier. The sound intensity was measured with a MPA-201 sound level meter. The total frequency range in the experiment was 100 Hz-2500 Hz. The acoustic pressure varied between 110 dB-128 dB. A fan was used to produce the flow field, and flow rate is fixed to 2 m^3/h (u0=0.0361 m/s). A steady flow plate is installed near the air-in to stabilized the airflow. The diameter of the waveguide tube is 14 cm, a soundproof sponge is installed at the end of the waveguide tube to avoid standing waves.

Copper spheres (λ=386 W·m^{-1}·K^{-1}, c_{ps}=384 J·kg^{-1}·K^{-1}, ρ=8440 kg·m^{-3}) with 0.005 m, 0.01 m and 0.015 m diameter were used for the heat-transfer studies. A hole was made in the center of the sphere, through which a miniature T-type thermocouple probe with a diameter of 5×10^{-4} and length of 0.04 m was inserted and positioned at the geometrical center of the sphere. Thermally conductive epoxy was used to stabilize the thermocouple in the hole. The spheres were positioned 0.02 m from the loudspeaker surface. The temperature measurements were performed after connecting the thermocouple to the data logger. The temperature data were logged every second for the sphere, and the initial temperature of the sphere was 120 °C prior the experiments. The thermocouples had a precision of up to two decimal values. The air supported by the fan was used as cooling medium, and its temperature was maintained at 27 °C. Experiments with flow field were performed as reference.

3.2. Results and discussion

In our experiment, a steady velocity field is formed in the waveguide tube by fan feeding. However, the air is injected into the tube through the air inlet, due to the pressure difference, a jet noise is formed. Figure 2 shows the spectrum distribution of the jet noise. It can be seen that the jet noise has a large sound pressure level at 500 Hz and 1450 Hz, respectively 112 dB and 109 dB.

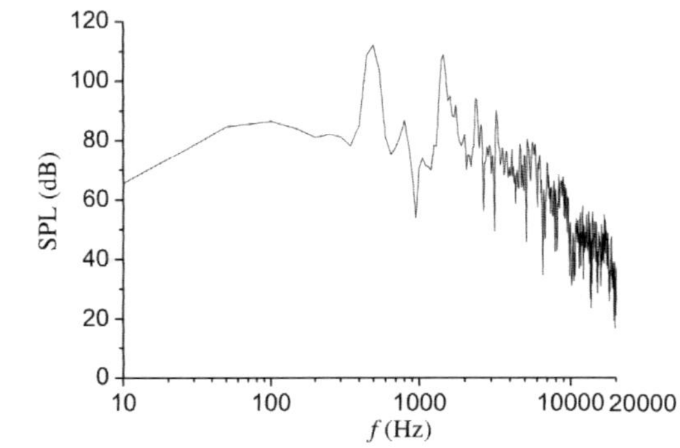

Figure 2. Spectrum distribution of jet noise

The sound pressure levels tested in our experiments were 110 dB, 115 dB, 120 dB, 125 dB and 128 dB. When the steady flow field (u_0=0.0361 m/s) is superimposed, the normalized curve of oscillating velocity is shown in the figure 3. And the velocity rate ($\varepsilon=u_1/u_0$, u_1 is the acoustic velocity) is 0.59, 1.07, 1.89, 3.38 and 4.76, respectively.

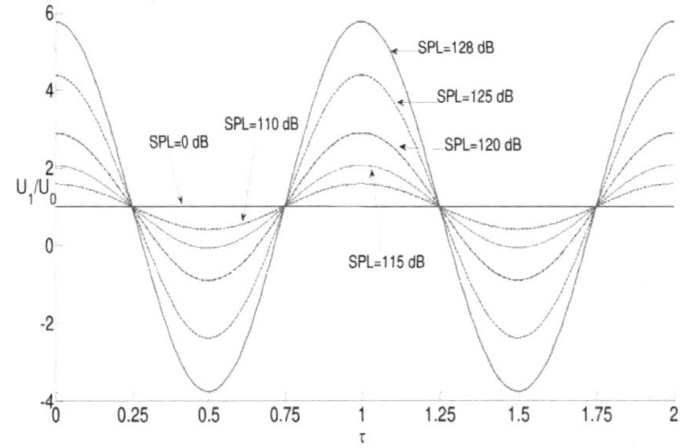

Figure 3. Normalized oscillating flow velocity in the presence of an acoustic field

Since acoustic radiation can generate heat, the amount of generated heat was also calculated following 1 min of irradiation after the sphere had reached equilibrium with the surrounding medium at 27 °C. The temperature increases were almost unmeasurable in the experiments, which means Q_S in Equation (6) or (7) can be ignored.

Figure 4 show the effect of an acoustic wave with a frequency of 500 Hz on the cooling curves of the copper spheres using the natural logarithm of temperature ratios (LnTR) as a dimensionless number according to the following equation, which can be expressed as:

$$\mathrm{LnTR} = \ln \frac{T - T_\infty}{T_i - T_\infty} \tag{10}$$

Each cooling curve in Figure 4 is an average of three repetitions. D is the diameter of the copper sphere. The smaller the value of LnTR after one minute, the quicker the cooling process was. As indicated in Figure 4, the shape of the LnTR curves is nearly linear, and the cooling rate increases with increasing intensity. However, when the diameter of the copper sphere is 5 mm and the sound pressure is 110 dB to 128 dB, the cooling rate are always lower compared to that without acoustic field. When the diameter of the copper sphere is 10 mm, the cooling rate are higher than that without acoustic field, as long as the velocity rate is larger than 3.38. As to the copper sphere with a diameter of 15 mm, the velocity rate is only need to be larger than 1.89.

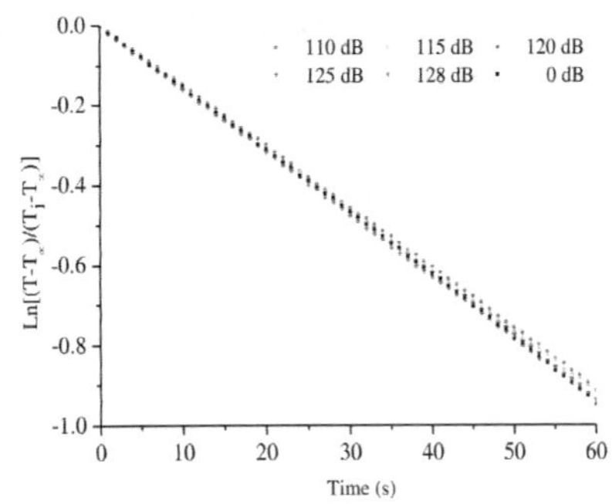

(a) Temperature gradient in 1 min when D=5 mm

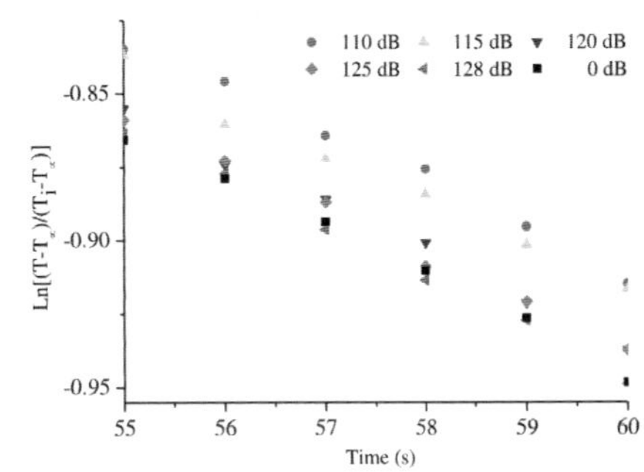

(b) Local magnification of temperature gradient in 5 seconds when D=5 mm

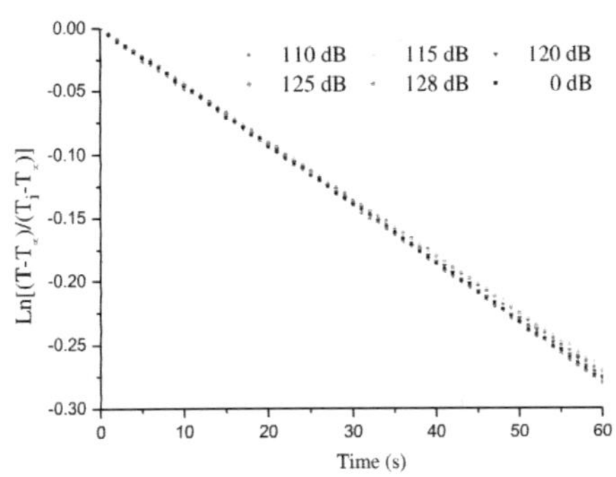

(c) Temperature gradient in 1 min when D=10 mm

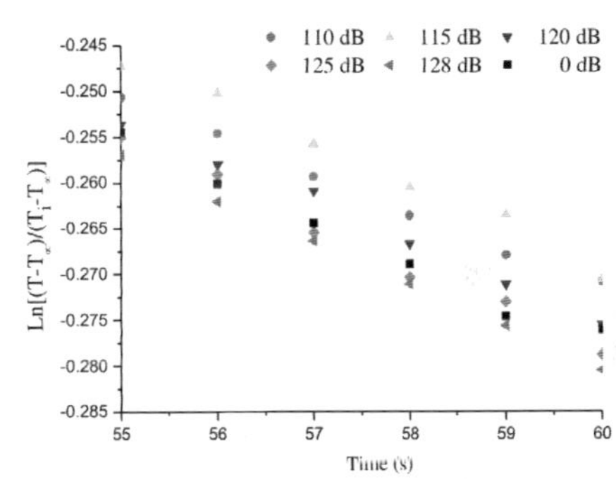

(d) Local magnification of temperature gradient in 5 seconds when D=10 mm

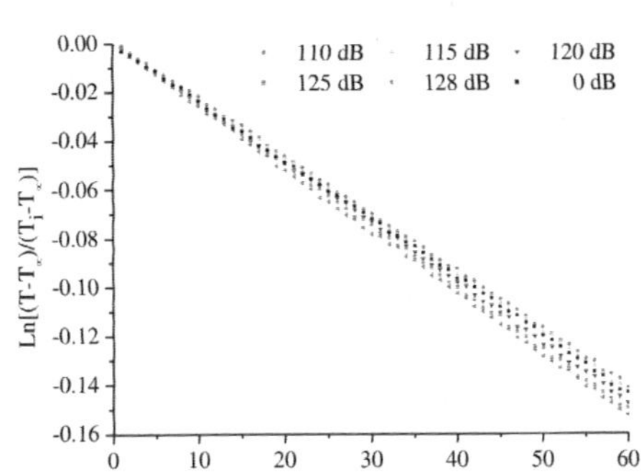

(e) Temperature gradient in 1 min when D=15 mm

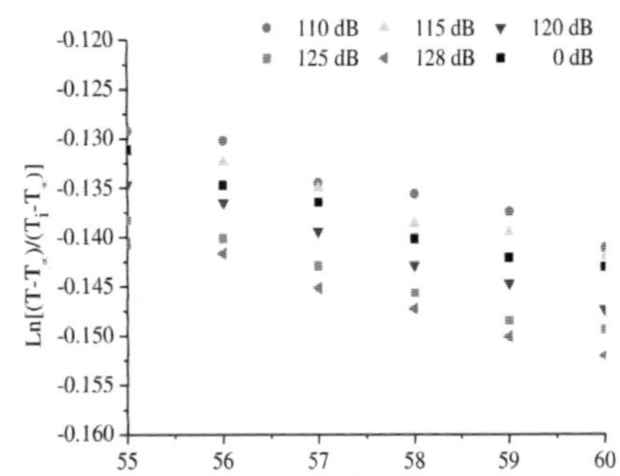

(f) Local magnification of temperature gradient in 5 seconds when D=15 mm

Figure 4. Effect of different SPL for frequency of 500 Hz on the excess temperature ratio during cooling a stationary sphere: U_0=0.0361 m/s

IOP Conf. Series: Materials Science and Engineering **626** (2019) 012014 doi:10.1088/1757-899X/626/1/012014

In Figure 5 the effect of different frequencies and the SPLs on the h value of the copper spheres is shown. The Eq. (7) describes the behaviour of the cooling curves very well and gives good fits with R^2 values between 0.998 and 0.9998. As we can see in Figure 5, the convective heat-transfer coefficient h is less than 50, so the Bi is far less than 0.1, which means the assumption in our experiment is reasonable and the temperature gradient across the copper sphere can be neglected.

Figure 6 shows the heat-transfer enhancement factor for different SPLs and frequencies. A larger Nu indicates a higher heat-transfer rate and higher enhancement factors. Therefore, figure 6 indicates that higher cooling rates of the sphere are achieved by increasing the intensity of the sound wave. The Nu values in figure 6 increase with increasing frequency at first, however when the sound frequency exceeded a certain frequency, the Nu values decrease with increasing frequency. And the specific frequency for the maximum value of Nu increases with increasing SPLs.

In addition, it can be seen in figure 6 that Nu decreases first and then gradually increases in the range of 100 Hz to 500 Hz, which may be caused by the obvious vibration of the waveguide tube, resulting in a large heat-transfer coefficient of the copper sphere.

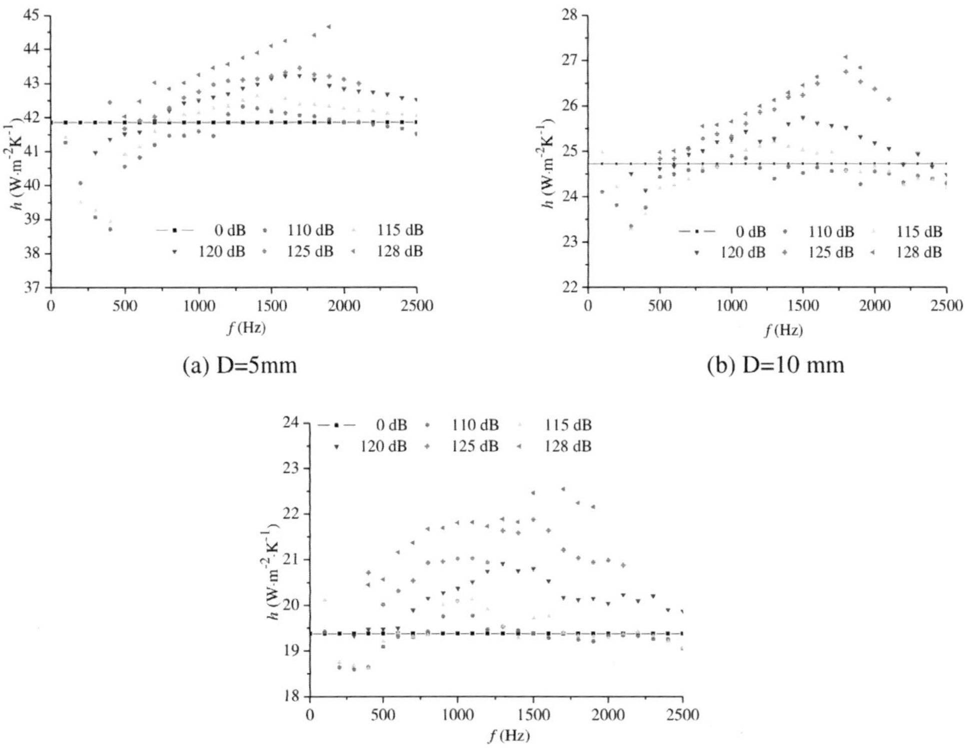

(a) D=5mm (b) D=10 mm

(c) D=15 mm

Figure 5. Distribution of convective heat-transfer coefficient h during cooling of a stationary sphere as a function of frequency: U_0=0.0361 m/s

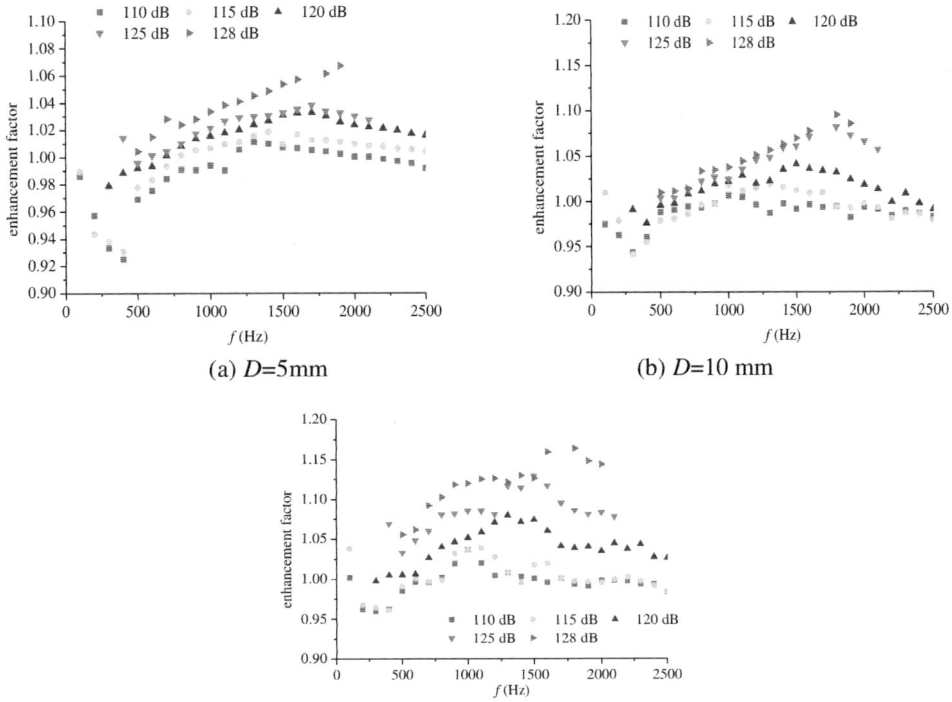

(a) D=5mm (b) D=10 mm

(c) D=15 mm

Figure 6. Distribution of heat-transfer enhancement factor during cooling of a stationary sphere as a function of frequency: U_0=0.0361 m/s

4. Conclusions

To investigate the heat-transfer characteristics of coal particles in the combined effect of acoustic field and flow field, an experiment was conducted. The following conclusions can be summarized:

(1) With the increasing of sound pressure level, the heat-transfer coefficient and heat-transfer enhancement factor of the copper spheres increase. However, when the diameter of the copper sphere is 5 mm, the cooling rate are always lower compared to that without acoustic field in our experiments. When the diameter of the copper sphere is 10 mm, the cooling rate are higher than that without acoustic field, as long as the velocity rate is larger than 3.38. As to the copper sphere with a diameter of 15 mm, the velocity rate is only need to be larger than 1.89. The larger the diameter of the copper sphere, the more prominent the acoustic effect;

(2) There is a specific frequency, at which *Nu* reaches a maximum. The specific frequency for the maximum value of *Nu* increases with increasing SPLs.

Acknowledgments

Support is acknowledged from the National Natural Science Foundation of China (51609226 and 11674093)

References

[1] Jiang Genshan, Zheng Yinchao, Pan Jie, Tian Jing. The enhancement of pulverized-coal combustion by using sound waves[J]. *The Journal of the Acoustical Society of America*, 2012, 131(4): 3468-3468

[2] Marthelli R C, Boelter L M. The effect of vibration on heat transfer by free convection from a horizontal cylinder[J]. *Proc. 5th Int. Congress of Applied Mechanics*, 1939, 578-584

[3] Lasen P S, Jense J W. Evaporation rates of drops in forced convection with superposed transverse sound field[J]. *Int. J. Heat Mass Transfer*, 1978, 21: 511-517

IOP Conf. Series: Materials Science and Engineering **626** (2019) 012014 doi:10.1088/1757-899X/626/1/012014

[4] Ha M Y, Yavukurt S. Combustion of a Single Carbon or Char Particle in the Presence of High Intensity Acoustic Field[J]. *Combustion and Flame*, 1991, 86: 33-46

[5] Gopinath A, Mills A F. Convective heat transfer from a sphere due to acoustic streaming[J]. *Journal of the Acoustical Society of America*, 1993, 94(3): 1773-1774

[6] Alassar R S, Badr H M, Mavromatis H A. Heat convection from a sphere placed in an oscillating free stream[J]. *International Journal of Heat and Mass Transfer*, 1999, 42(7): 1289-1304

[7] Kiani H, Sun D W, Zhang Z. The effect of ultrasound irradiation on the convective heat transfer rate during immersion cooling of a stationary sphere[J]. *Ultrason Sonochem*, 2012, 19(6): 1238-1245

[8] Kiani H, Sun D W, Zhang Z. Effects of processing parameters on the convective heat transfer rate during ultrasound assisted low temperature immersion treatment of a stationary sphere[J]. *Journal of Food Engineering*, 2013, 115(3): 384-390

[9] Weilong Xu, Genshan Jiang, Liansuo An, Yuechao Liu. Numerical and experimental study of acoustically enhanced heat transfer from a single particle in flue gas[J]. *Combustion Science and Technology*, 2018, 190(7): 1158-1177

Experimental investigation of elastic wave propagation and inertia effects in extra soft gelatin material

H.B. Zeng, P. Bailly

LaMé, INSA CVL, Univ. Orléans, Univ. Tours, F-18022 Bourges, France

huabin.zeng@insa-cvl.fr

Abstract. In this paper, elastic wave propagation and inertia effects in gelatin-based soft material under impact loading is investigated. The dynamic test is principally performed by the classical SHPB (Split Hopkinson Pressure Bars) technique with highly sensitive piezoelectric polyvinylidene fluoride (PVDF) pressure sensors placed on the interfaces between the specimen and the bars. Digital image correlation (DIC) technique is conducted to measure the strong non-homogeneous strain field, the particle velocity, the the strain rate, the strain acceleration under transient stress state using high speed photography. The particle velocity and its gradient are used to extract the elastic wave velocity. Additionally, we derive closed-form equations for the additional axial stress produced by transverse inertia in the rectangular parallelepiped specimen under the non-equilibrium stress state. The amplitude of transverse inertial stress depends strongly on the strain rate and the strain acceleration. The transverse inertia results in a significant spike-like stress in the initial portion of inertial stress history and in the zone of the specimen near the impact side due to the acceleration stage of the impact test. It is obvious that the degree of inertia effects depends strongly on time steps and spatial positions.

1. Introduction

The extra soft gelatin-based materials are widely used in the biomechanical, civilian and military application due to the low stiffness, mechanical impedance, wave velocity and sensitivities to strain-rates. However, these characteristics lead to some difficulties in experimental measurement to obtain the dynamic behaviour under impact loading [1-4]. Split Hopkinson pressure bar is used for high strain rate characterization of ballistic gelatin [5-6] and previous theories [7-8] discussed many equations that predict the radial inertia effects on cylindrical gelatin specimens under dynamic force equilibrium loading. Recent developments in the domain of optical full-field measurements, especially the digital image correlation (DIC) is investigated to measure the non-homogeneous strain and stress filed in soft materials [9-11]. This type of non-parametric method was successfully developed by the authors of the present work to characterize dynamic behaviour of incompressible soft gelatin-based specimens in SHPB tests under non-equilibrium force state [12]. In this paper, the particle velocity field and elastic wave speed are analysed by using DIC and we derive closed-form equations for the additional axial stress produced by transverse inertia in the rectangular parallelepiped specimen under the transient non-equilibrium stress state during the SHPB impact tests.

1.1. Material and specimen geometry

The gelatin specimen examined in this investigation is based on SEBS (Styrene-ethylene/butylene-styrene) material. The density of gelatin material and compressive elastic modulus were measured and found to be 967 kg/m3 and 250 kPa respectively. The target material specimen is presented in Figure 1,

IOP Conf. Series: Materials Science and Engineering **626** (2019) 012015 doi:10.1088/1757-899X/626/1/012015

exhibiting rectangular parallelepiped with 22 mm length in e_1 direction, 14 mm height in e_2 direction and 12 mm thickness. In order to facilitate the use of digital image correlation, a high contrast random speckle pattern is applied on the lateral surface of specimen. The speckle pattern consists of random white and black particles and it was carried out using normal aerosol paint.

1.2. Dynamic impact loading

A specific nylon split Hopkinson pressure bar apparatus is used as dynamic test method to determine the dynamic behavior of low impedance materials at high strain rates [14-15]. The SHPB system consists principally of three parts: the projectile bar with 1.5mm length, the lengths of the incident bar and transmitted bar were 3.00 m and 1.75 m respectively (Figure 2). The bars possess the same diameter (20 mm). Two highly sensitive piezoelectric PVDF pressure sensors are applied to access a valid stress measurement because of extra low mechanic impedance of gelatin. One PVDF sensor is sandwiched between the input bar and the specimen to measure stress at the impact side (Figure 2), another PVDF sensor is placed between the specimen and output bar to capture the stress developed on the bottom end of the specimen.

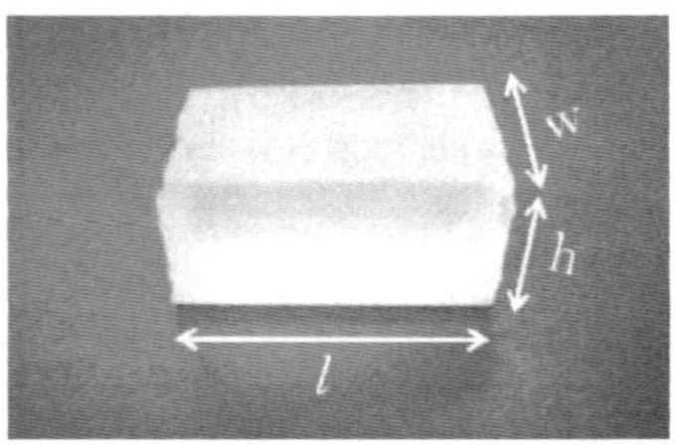

Figure 1. Photograph of gelatin specimen SEBS

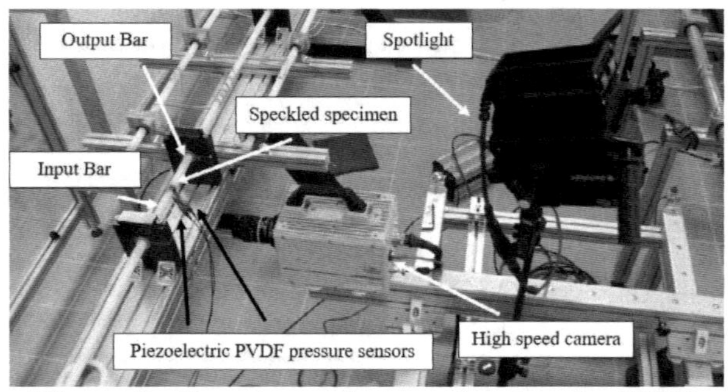

Figure 2. Experimental setup showing the SHPB apparatus with two PVDF sensors.

1.3. Digital image correlation

The lateral face of the specimen is speckled and the images of the test are recorded by a high speed camera PHOTRON. The digital image correlation (DIC) technique is used to access the lagrangian displacement *U(X,t)* in this work. The images of the speckled lateral face of the specimen are recorded during the tests by a PHOTRON high speed camera. The Figure 3 presents the field of axial strain readily extracted from DIC analysis at an instant during the test under impact velocity 2.9 m/s. The inhomogeneous strain field varies according to the axial X_1 direction.

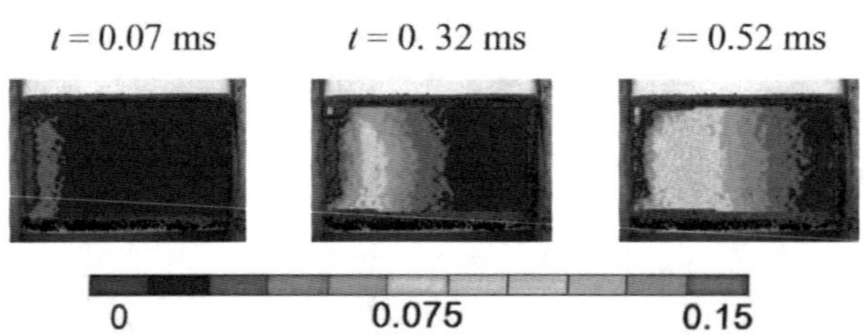

Figure 3. Axial strain on the lateral side of gelatin specimen at different time steps.

2. Data analysis

2.1. Axial stress

The linear momentum conservation in Lagrangian equation in the specimen can be written as:

$$\underline{div}\ \underline{\underline{S}}(\underline{X},t) = \rho \ddot{\underline{U}}(\underline{X},t) \tag{1}$$

IOP Conf. Series: Materials Science and Engineering **626** (2019) 012015 doi:10.1088/1757-899X/626/1/012015

Where S denotes the « nominal stress » (Polia-Kirchhoff 1), U is the lagrangian displacement filed and ρ is the material density.

The evolution of average axial stress in the cross section is included by using the average acceleration of the section, as

$$\frac{\partial S_{11}(X_1,t)}{\partial X_1} = \rho \ddot{U}_1(X_1,t) \tag{2}$$

Distribution of transient axial stress in time and space can be obtained by applying the measured stress history on the input face, $S_{11}(0,t)$, by the PVDF pressure sensor and the axial acceleration data extracted from DIC in the axial inertial stress item, as

$$S_{11}(X_1,t) = S_{11}(0,t) + \int_0^{X_1} \rho \ddot{U}_1(X_1,t) \tag{3}$$

2.2. Transverse inertial stress

From Nishida [7] and Song [8], in our Cartesian coordinate's notation their equation for the transient axial stress component in e_1 direction is given by

$$S_{11}(X_1,t) = S_{11}^0(X_1,t) + P(X_1,t) \tag{4}$$

where S_{11}^0 denotes the axial intrinsic stress, P is the transverse inertia stress and t is time.
From the equation of motion in the transverse direction in the rectangular parallelepiped specimen under the transient non-equilibrium stress state, we derive the equation for P as

$$P(X_1,t) = \left(\frac{\rho(\dot{\varepsilon}_{11}(X_1,t))^2}{8(1-\varepsilon_{11}(X_1,t))^3} + \frac{\rho\ddot{\varepsilon}_{11}(X_1,t)}{12(1-\varepsilon_{11}(X_1,t))^2}\right)(H_o^2 + W_o^2) \tag{5}$$

where ε_{11} denotes the axial strain, $\dot{\varepsilon}_{11}$ and $\ddot{\varepsilon}_{11}$ are the strain rate and strain acceleration respectively and they can be extracted from the axial strain field obtained by DIC. H_o and W_o represent half the height and half the width of the specimen respectively.

3. Results and discussion

3.1. Particle velocity and wave speed

Once the establishment of the contact between the input bar and the specimen, an elastic stress wave is immediately developed at the impact side of the specimen and propagates toward the bottom end. The particle velocity profile can be deduced from the lagrangian displacement *U(X,t)* obtained by DIC (Figure 4). The elastic wave front can be tracked at each time step by carrying out the axial location of the maximum value of the particle velocity gradient (max dV_p/dX_1). For example, at t=0.15ms, the axial location of maximum velocity gradient can be found as X_1=4.4 mm in Figure 4 and The elastic wave front reaches at this location at the moment. The elastic wave speed of the gelatin specimen is extracted by plotting the variation of the location of elastic wave front in time in Figure 5. One can calculate the elastic wave speed from the slope of the best fitted linear regression line. Consequently, the elastic wave speed was deduced as 18 m/s. In addition, the theoretical elastic wave speed by assuming a uniaxial stress condition was obtained as 17.5±1.5 m/s.

3.2. Inertial stress

Figure 6 shows that transient axial stress and transverse inertial stress along the specimen at different times under impact velocity 2.9 m/s. It is observed that during the dynamic non-equilibrium test, the degree of inertia effect depends strongly on time steps and spatial positions for the extra-soft gelatin-like materials. At time step 0.22 ms, the transverse inertia stress is thus significant in the initial loading portion and the inertia stress peak occurs in the zone of the specimen near the impact face. In this case the strain rate need to takes time to increase from zero to the high value so that it make the initial inertia acceleration marked. The inertial stress decreases steeply in the bottom direction due to the low wave speed of gelatin. For t=0.67 and 0.97 ms, the inertial stress became relatively important in the middle ant the bottom part owing to the particle movement reduced and stopped by the output bar respectively.

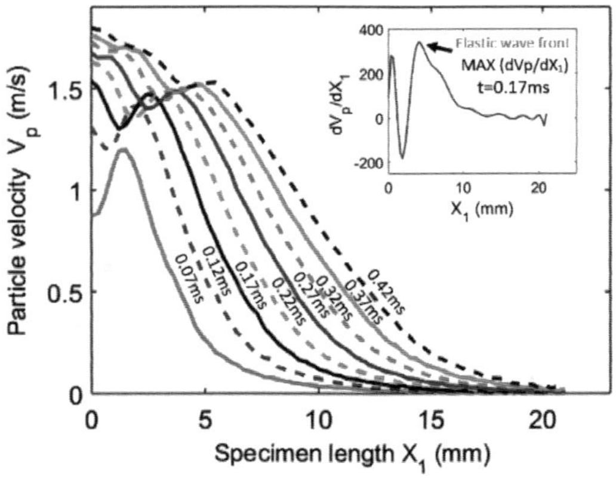

Figure 4. Spatial variation of particle velocity at different times.

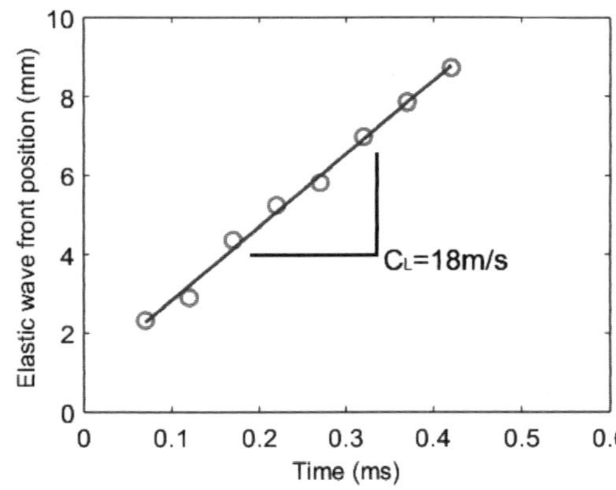

Figure 5. Elastic wave front speed calculation.

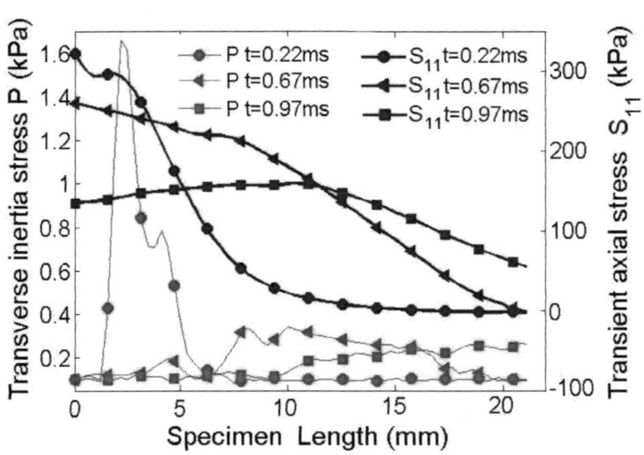

Figure 6. Transient axial stress and inertia stress along the specimen.

4. Conclusion and perspectives

In this work, an experimental setup based on the DIC technique is developed to investigate the elastic wave propagation and inertial effect in extra low impedance gelatin material using high speed photography. The SHPB apparatus is used for dynamic compression testing on rectangular parallelepiped specimen. The weak impact forces on the extra soft gelatin specimen is captured by highly sensitive PVDF pressure sensors. A strong non-homogeneous strain field, the particle velocity, the strain rate and strain acceleration are extracted from the DIC measurement by the images captured by high speed photography. The experimentally deduced elastic wave velocity and the theoretically calculated longitudinal sound velocity are consistent. In addition, the closed-form equations is proposed for the transverse inertia in the rectangular parallelepiped specimen under the transient stress state during

impact test. It should be emphasized here that the transverse inertia in specimen during dynamic test leads to an additional axial stress in the form of spike in the initial portion of the transverse inertial stress and in the zone of the specimen near the impact face. The amplitude of transverse inertial stress depends on the strain rate, the strain acceleration and the geometry of specimen.
Future work concerns interface friction analysis between incident/transmitted bars and specimen by DIC investigation.

Acknowledgements

The authors recognize and acknowledge the LaMé Laboratoire (Laboratoire de Mécanique Gabriel Lamé) for the financial support of the presented work.

References

[1] Pervin F, Chen W 2009 Dynamic mechanical response of bovine gray matter and white matter brain tissues under compression. J Biomech 42:731–735
[2] Song B, Chen W 2004 Dynamic stress equilibration in split Hopkinson pressure bar tests on soft materials. Exp Mech 44:300–312
[3] Chen W, Zhang B, Forrestal MJ 1999 A split Hopkinson bar technique for low-impedance materials. Exp Mech 39:81–85
[4] Song B, Ge Y, Chen W, Weerasooriya T 2007 Dynamic and quasi-static compressive response of a porcine muscle. J Biomech 40:2999–3005
[5] Chen, W., Zhang, B., Forrestal, M.J., 1999. A split Hopkinson bar technique for lowimpedance materials. Exp. Mech. 39, 81–85.
[6] Kwon, J., Subhash, G., 2010. Compressive strain rate sensitivity of ballistic gelatin. J. Biomech. 43, 420–425.
[7] Nishida, E and Chen W.W., 2010. "Inertia effects in the impact response of soft materials", in Proceedings IMPLAST 2010 Conference, publ. Bethel, CT, Society for Experimental Mechanics
[8] Song B, Ge Y, Chen W, Weerasooriya T., 2007. Radial inertia effects in Kolsky bar testing of extra-soft specimens. ExpMech 47:659–670
[9] H. Khatam · Q. Liu · K. Ravi-Chandar. 2014 Dynamic tensile characterization of pig skin. Acta Mechanica Sinica;30(2) :125–132.
[10] R. Othman, S. Aloui, A. Poitou, 2010 Identification of non-homogeneous stress fields in dynamic experiments with a non-parametric method, Polymer Testing;29: 616 – 623.
[11] B. Koohbor, A. Kidane, W.Y. Lu, M. A. Sutton. 2016 Investigation of the dynamic stress–strain response of compressible polymeric foam using a non-parametric analysis. International Journal of Impact Engineering;91:170–182.
[12] Zeng H.B., Bailly P., 2017. Experimental characterization of dynamic behavior of gelatin-based material using DIC. Polymer Testing 63: 298-306
[13] H. Zhao, G. Gary, 1995 A three dimensional analytical solution of the longitudinal wave propagation in an infinite linear viscoelastic cylindrical bar. Application to experimental techniques, Journal of the Mechanics and Physics of Solids 43 (8) 1335 –1348.
[14] H. Zhao, G. Gary, J. Klepaczko, 1997 On the use of a viscoelastic split Hopkinson pressure bar, International Journal of Impact Engineering 19 (4) 319–330.

IOP Conf. Series: Materials Science and Engineering **626** (2019) 012016 doi:10.1088/1757-899X/626/1/012016

Effect of Abutment Design Parameter on Mechanical Properties for Dental Implant System

Xiuyuan Lu[1,2], Xiao Zhang[1,2,3], Jianyu Chen[2] and Xianshuai Chen[2]

[1] Guangzhou Janus Biotechnology Co.,Ltd., Guangzhou, 511458, PR. China
[2] Foshan Angels Biotechnology Co.,Ltd., Foshan, 528000, PR. China
[3] Author to whom any correspondence should be addressed.

xiao.zhang@gzjanus.com

Abstract. Abutments, as the intermediate connection between implant and crown vary on external characters decided by design parameters in order to suit for multiple clinical cases with the development of prosthodontics. The various angle, groove height (GH) and diameter of abutment brought different magnitude of stress to the dental implant system. Excessive stress concentration lead to fatigue fracture failure. Thus, in this study, parametric design and Finite Element Analysis (FEA) were combined to understand the effects of angled-abutment GH and diameter on the mechanical properties of the implant system. 30 models of abutment with varying angles (15° and 25°), diameters (4, 5 and 6mm) and GH (1, 2, 3, 4 and 5mm) were simulated and the values of Mises the equivalent stresses were analyzed by two-way ANOVA. The results demonstrated that the three design parameters had no influence on the area where stress concentration occurred and the diameter had no effect on the maximum stress while the GH had a significant influence ($P<0.05$).

1. Introduction

The dental implant-abutment system has been gradually approved and widely used to improve the quality of patients' life over the past 40 years since the osseointegration was discovered by Prof. Brånemark of Sweden [1]. The planting position varies from case to case in contemporary implant dentistry, traditional straight abutments do not perfectly solve all clinical cases, which requires more specifications and types of abutments to connect implants and crowns. At present, abutment was shaped by three main parameters: angle, diameter and GH.

Long-terms of stability and survival rate of implant system rely on its outstanding mechanics performance. In the process of swallowing, chewing and speaking, the implant system in mouth experienced different levels of load. Moreover, different geometric shapes of the abutments produce various levels of stress concentration [2], which may lead to cracks, and accelerate crack propagation [3]. Kao HC et al [4] demonstrated that the 25°-angled abutment increased the stress level of the surrounding cortical bone and displacement of the implant by using FEA. Clelland's study [5-7] showed that the maximum stress of the abutment with angle of 15° and 20° were 170% and 190% compared with the straight abutment; Kebin Tian et al [8] evaluated the influence of the titanium alloy angled abutments' geometry on the stress distribution by means of FEA and the results showed that angled abutments resulted in decreased stresses when implants were not placed in ideal axial position.

However, little research has been focused on how the GH and diameter affects the stress level of the implant system. Therefore, a virtual prototype needs to be built and simulated to understand the influence of different design parameters on the mechanical properties of the whole implant system.

IOP Conf. Series: Materials Science and Engineering **626** (2019) 012016 doi:10.1088/1757-899X/626/1/012016

This paper focuses on the effects of design parameters of the abutment on the stress level and distribution of the implant system.

2. Material and methods

2.1. 3D models

The dental implant-abutment system consisted of implant body, central screw and abutment. It was proved that the smallest diameter of implant (D=3.5mm) and central screw (M1.8) caused the maximum stress [9]. In the meanwhile, implants with same diameter but different length (L) showed little effect on the stress level [10], Therefore, the implants (D=3.5mm, L=10mm) and the central screws (M1.8) were used in this study in order to amplify the stress concentration under extreme conditions.

The implant body and central screw was simplified to cylinder in the case that all the thread connections were ideal and valid to focus on the regular of stress for some study have suggested that cylindrical implant models might be a good approximation of screw-type implants in further FEA [11].

2.2. Experimental design

The abutment GH was the sum of the depth of the gum combined with the length of the epithelium and attachment of the connective tissue, additionally groove surface needed to fit the surgical wound of gum. Therefore, the ideal GH was 2.0-5.0 mm [12] and ideal diameter was 4-6mm, and generally abutment angle ranged from 0-30° on the basis of surgical plan. In order to get detailed data, 30 models were divided into to two parallel groups established at 15° and 25°. The GH specifications of each group were 1mm, 2mm, 3mm, 4mm, and 5mm; the diameter specifications were 4mm, 5mm, and 6mm.

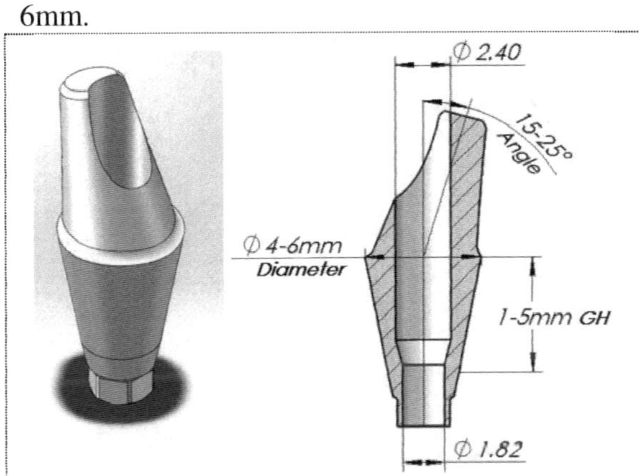

Figure 1. 3D models of abutment.

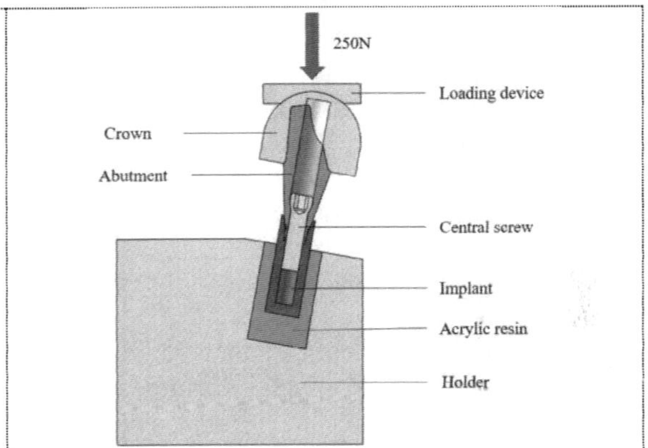

Figure 2. Complete 3D model for EFA

2.3. Parameterization process

By using the parameterized module of Ansys 16.0 Workbench, the solid model modified by Solidworks and geometry used in finite element analysis were connected. Prefix "DS_" needs to be added ahead the name of the dimensions which separately decide the diameter, GH and angle of the abutment. This process allows all models to be calculated all at once without resetting boundary conditions and contact condition. Solid model and parameters range are shown in Figure 1.

2.4. Finite element analysis (FEA)

A standardized implant fatigue testing method (ISO 14801) [13] declares a detailed Implant system test model as shown in Figure.2. Material properties were shown in the Table 1. All materials were considered to be isotropic, homogeneous, and linearly elastic [14]. In this study, assembly was meshed with tetrahedral elements, which composed of 361010-495010 element and 240756-338899 nodes.

IOP Conf. Series: Materials Science and Engineering **626** (2019) 012016 doi:10.1088/1757-899X/626/1/012016

Table 1. Material properties for FEA [9]

Material	Young's modulus (Pa)	Poisson ratio	Tensile Yield Strength (Pa)	Compressive Yield Strength (Pa)	Assignments
TC4	9.6E+10	0.36	9.3E+8	9.3E+8	Crown, abutment, central screw
GR4	9.6E+10	0.36	4.9E+8	4.9E+8	Implant
Acrylic resin	3.78E+9	0.36		54.6	

Completely fixation was applied to thread contact surfaces [15], and liner connection was applied to the remaining interfaces. Additionally, a force of 250N (average bite force of adult people) [16, 17] with vertically downward direction was applied to the crown.

The final results of the finite element analysis were evaluated by the maximum Mises equivalent stresses (σ_{VM}). Stress concentration in the stress cloud map revealed the most vulnerable area in the model [18]. Two-way analysis of variance (Two-Way ANOVA) was performed on two parameters -the diameter and the GH of abutment. The significance level was set at 5%. This procedure allowed the authors to calculate the contribution of each parameters and find out the main effects.

3. Results & Discussion

The stress cloud diagrams revealed that stress distribution was independent of design parameters of abutment. The peak value of σ_{VM} all appeared on abutment. Implant system with the 25°-angled abutment (GH=5mm, D=6mm) was taken as an example in Figure 3.

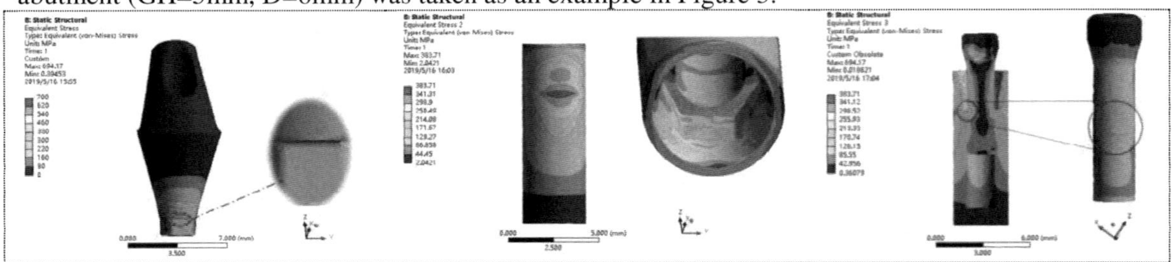

Figure 3. The stress distribution diagrams of abutment, implant body and central screw.

The σ_{VM} =690.3 MPa was distributed at the rounded corners of the hexagonal joint where the abutment shape change was most severe. Therefore, it could be speculated as the dangerous section.

The stress distribution of implant illustrated the neck region where implant body and the upper surface of acrylic resign material contacted as well as the inner surface contacted with the region below groove surface had a higher stress concentration. The σ_{VM} = 383.71MPa was much lower than abutment.

For central screw, the stress mainly concentrated at the thread surface fitted with the first two inner thread of the implant. Since the two regions appeared same level stress concentration on the same cross section, fatigue fracture was quite likely to occur under cyclic stress.

IOP Conf. Series: Materials Science and Engineering **626** (2019) 012016 doi:10.1088/1757-899X/626/1/012016

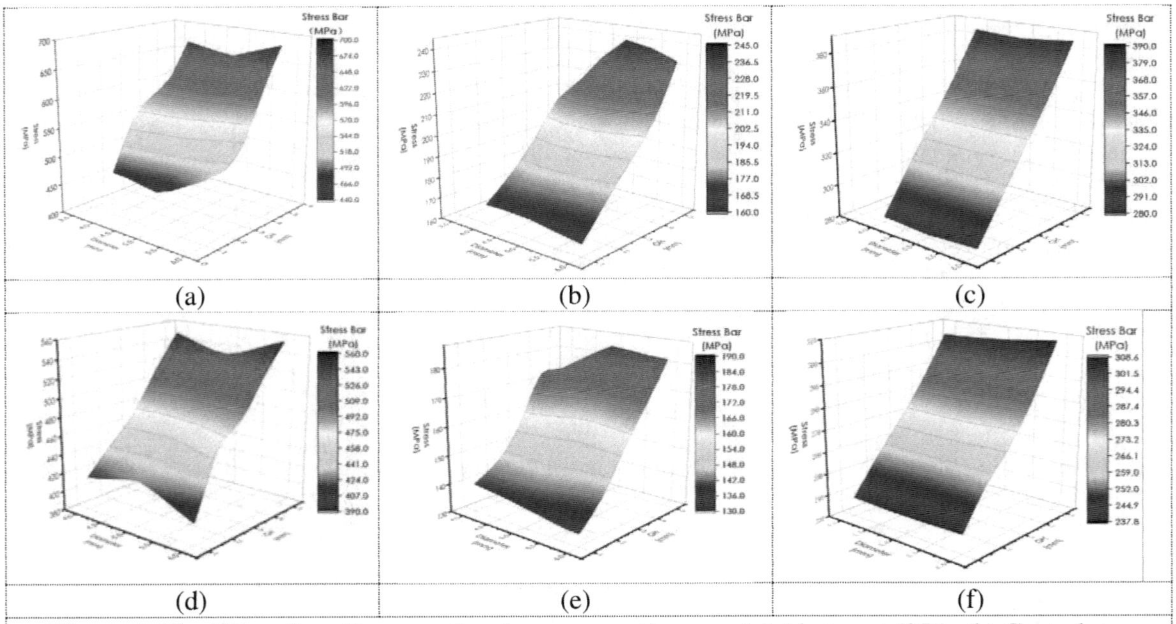

Figure 4. The 3D surface diagram of each component's σ_{VM}: (a) Abutment (25°); (b) Central screw (25°); (c) Implant (25°); (d) Abutment (15°); (e) Central screw (15°); (f) Implant (15°).

As shown in Figure 4, for the 25°- angled abutment, the peak value of σ_{VM} was 690.3 MPa (GH=5mm) while the nadir was 458.4 MPa (GH= 1mm). And for the 15°- angled abutment, the peak was 555.4 MPa (GH=5mm) while the nadir was 414 MPa (GH=1mm).

Apparently, the σ_{VM} of implant body, central screw and abutment increased with GH increasing when abutment has certain angle (15° or 25°). The contour lines in the 3D surface figure were almost parallel to the diameter axis, which means the dimeter of abutment was foreign to σ_{VM}, and it was confirmed by two-way ANOVA. For the implant system with 15°- angled abutment, the GH had a significant influence on the equivalent stress of the abutment ($F=26.82$; $P=0.035<0.05$) while the diameter had no influence ($F=0.29$, $P=0.65$). The same result appeared when the angle was 25° that the GH has a significant influence on the equivalent stress of the abutment ($F=36.37$; $P=0.026<0.05$) while diameter has no impact ($F=0.068$, $P=0.82$).

In the case where the threaded connection (central screw-implant, implant-resin) was always valid, abutments' GH had various extent of impact on the three primary components. Implant and abutment were less affected than the central screw. the σ_{VM} of Central screws in 25° group was 130% compared with the counterparts in 15° group while the implant and abutment were 125%.

Theoretically, abutment with short GH and tiny angle brings optimistic effect of oral rehabilitation based on the above analysis. However, GH was decided by actual situation of clinical cases. This study suggests that if the abutment GH has to maintain the worse condition, angle and/or diameter of implant and central screw need to be reconsidered in surgical plan to decrease the stress level.

4. Conclusion

The aim of this study was to understand how abutments' three parameters affect the stress distribution among the 30 abutment models. The following conclusions were drawn: stress distribution was independent of the three design parameters of abutment; The maximum stress appeared at the abutment with longest GH and bigger angle; Besides, abutment GH had significant influence on stress level while diameter had no influence; Abutment angle aggravated the influence of GH by comparison. In the further study, systematically contradistinction of GH and angle's impact will be made to drawn a rigorous conclusion.

Acknowledgement
This work was supported by 1) the Science and Technology Program of Nansha District, Guangzhou (Technology Development Program, No. 2017KF007); 2) the Science and Technology Program of Guangzhou (No. 201906010032);

References
[1] Branemark P I, Adell R and Breine U 1969 *Scand J Plast Reconstr Surg.* **3** 81.
[2] Hasan B, Roger F, Heinemann L and Keilig C 2012 *Medical Engineering & Physics* **34** 817.
[3] Cornell K, L E E, Matthias K, J and Robert Kelly 2009 *Entail Materials.* **25** 1419.
[4] Kao H C, Gung Y W, Chung T F and Hsu M L 2008 *Int J Oral Maxillofac Implants.* **23** 623.
[5] Clelland N L, Lee J K and Bimbenet O C 1995 *J prosthodont.* **4** 95.
[6] Clelland N L and Gilat A 1992 *J Prosthodont.* **1** 24.
[7] Clelland N L, Gilat A. McGlumphy EA and Brantley WA 1993 *Int J Oral Maxillofac implants.* **8** 541.
[8] Kebin T, Jiang C, Lilin H, Jin Y, Weniu H and Dong W 2012 *Medical Engineering & physics.* **34** 1526.
[9] Jinyang Z, Maisheng L, Xiao Z and Xianshuai C 2019 *International Journal of Materials. Mechanics and Manufacturing.* **7** 105.
[10] Lucie H, Tat'jana D and Alois K 2004 *The journal of prosthetic dentistry.* **91** 20.
[11] Okumura N, Stegaroiu R and E. Kitamura 2010. *Journal of Prosthodontic Research.* **54** 133.
[12] Li-na W and Zhen F 2017 *Journal of oral and maxillofacial surgery.* **27** 290.
[13] *ISO 14801:2016 Dentistry —Implants —Dynamic fatigue test for endosseous dental implants.*
[14] Tao L, Kaijin H, Libo C, Yin D and Yuxiang D 2011 *Applied Mathematical Modelling.* **35** 446.
[15] Amer M M, Rashad H A and Abdallah S 2015 *Tanta Dental Journal.* **12** 30.
[16] Sushant M and Raosaheb P 2016 *International Research Journal of Engineering and Tec hnology.* **2** 850.
[17] Lang L A, Kang B and Wang R F 2003 *Prostehet Dent.* **90** 539.
[18] Eraslan O, Sevimay M, Usumez A and Eskitascioglu G 2005 *J Oral Rehab.* **32** 273.

IOP Conf. Series: Materials Science and Engineering **626** (2019) 012017 doi:10.1088/1757-899X/626/1/012017

Numerical Analysis of Influence of Welding Sequence on Welding Deformation of Magnetic Separator Dielectric Box

Xuan Sun[a*],Zhiyong Wang[b], Li Yingjun and Ai Changsheng

School of Mechanical Engineering, University of Jinan, Jinan, China

[a] me_sunx@ujn.edu.cn, [b] 971917165@qq.com

Abstract. Aiming at the problem that the welding sequence has a great influence on the welding deformation of the dielectric box, this paper uses the nonlinear finite element method to numerically simulate the four welding sequences of the magnetic separator dielectric box, and analyzes the transverse deformation and longitudinal deformation. The welding sequence with the smallest amount of deformation was obtained, and the correctness of the simulation results was verified by experiments.

1. Introduction

The magnetic separator is an important iron ore screening equipment in the iron ore smelting industry, which can screen iron powder from crushed iron ore. The dielectric box is one of the most important parts of the magnetic separator, which determines the purity and efficiency of the iron powder screening. The welding of the dielectric box is to weld the two ends of the soft magnetic rod to the surface of the soft magnetic alloy plate, as shown in Figure 1(b).

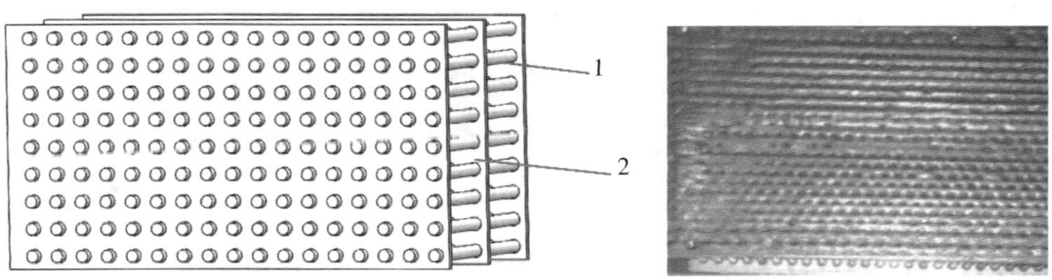

(a)Before the dielectric box is welded(b)After the dielectric box is welded
1-the soft magnetic rod 2-the soft magnetic alloy plate

Figure 1. Magnetic separator dielectric box

In the multi-pass welding process, due to the locality and non-uniformity of the temperature field[1], residual stress and welding deformation of the dielectric box weld bead are formed, which affects the dimensional accuracy of the dielectric box, and thus affects the assembly relationship with other parts. For multi-pass welding, different welding sequences have different effects on residual stress and deformation during the welding process. If the welding sequence is not properly selected, the residual stress after welding will be large and the amount of welding deformation will be large[2]. It is of great significance to numerically simulate the welding deformation of different welding sequences by numerical simulation technology, and to select the optimal welding sequence.

2. Welding finite element model establishment

2.1.Grid partition

Since the weld bead are concentrated on the surface of the alloy plate during the welding process, the welding deformation is mainly concentrated on the alloy plate, and the assembly accuracy of the dielectric box is mainly determined by the dimensional accuracy of the alloy plate. Therefore, in order to simplify the analysis model and improve the efficiency of the solution, the welding simulation of the dielectric box is simplified to the multi-pass welding simulation of the thin plate. Study the effect of different welding sequences on the welding deformation of dielectric box alloy plates. Firstly, a finite element model of 10 weld beads with a thickness of 5mm was established by using finite element simulation analysis technology. In the process of dielectric box welding simulation, in order to further improve the efficiency of the solution, the following assumptions were made:

1) The surrounding environment of the welding process and the initial temperature of the dielectric box is 20℃;

2) Ignoring the difference between the soft magnetic alloy plate and the soft magnetic bar, unified into the same isotropic material;

3) The movement speed of the welding torch during the welding process is constant, and the heat generated by the arc strictly obeys the Gaussian heat distribution;

4) Ignore the heat transfer between the dielectric box and the fixture, only consider the convective heat transfer and radiative heat transfer between the media box and the surrounding environment.

The weld bead formed by the welding of the dielectric box is shown in Figure 1(b), and the weld bead is a smooth straight bead. Therefore, the weld bead can be set to a long rectangular shape during the welding simulation, and the weld bead size is 400mm long, 5mm wide and 5mm thick. The soft magnetic alloy plate is 400mm long, 150mm wide and 5mm thick. In order to reduce the calculation amount and simulation complexity, and to achieve the purpose of this simulation test, this paper only simulates the soft magnetic alloy plate. The welding stress and welding deformation are studied. The geometric model of the alloy plate is shown in Figure 2(a).

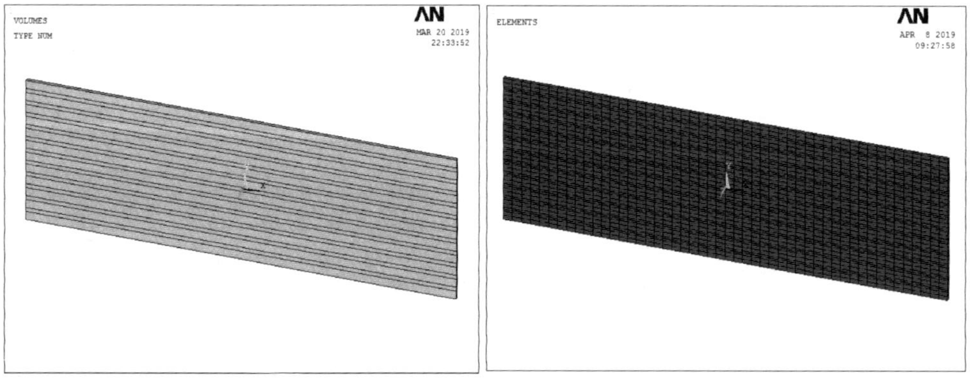

(a)The alloy plate geometry model (b)Results of alloy plate meshing

Figure 2. Geometric model and meshing of dielectric box alloy plates

In the meshing, a smaller structured grid is used for the melting zone and the heat-affected zone of the weld bead, and a sparse structured grid is adopted for the non-heat-affected zone far from the bead[3]. The result of the meshing is shown in the Figure. 2(b).

2.2. Material thermodynamic parameters

The alloy plate is a nickel-iron alloy, and its physical properties and mechanical properties change with temperature. For performance parameters above 800 °C, the difference between inside and outside the law is used[4].Meanwhile, in order to simplify the calculation, the bead unit and the base metal have the same thermophysical parameters. The variation of each parameter with temperature is shown in Table 1.

2.3. Boundary condition setting

When performing thermal-structural coupling analysis of the weld, it is generally necessary to define the structural constraints and the heat transfer boundary conditions[5].The structural constraint is defined to prevent the workpiece from producing a rigid displacement during the analysis process, and at the same time, it cannot affect the stress release and free deformation during the welding process.Therefore, the joints at both ends of the alloy plate 5mm from the weld bead. were restrained. The heat transfer boundary conditions generally include heat conduction, convection heat transfer and radiation heat transfer.Neglecting the contact heat transfer between the workpiece and the fixture, only considering the convective heat transfer and radiation heat transfer of the outer surface of the workpiece. Where, the convection heat transfer coefficient between the air and the workpiece is 15W/(m^2·K), and the radiative heat transfer coefficient within each temperature range is calculated according to Newton's cooling formula, and is added with the convection heat transfer coefficient, which is combined to form the surface composite heat transfer coefficient.

Table 1. Thermodynamic parameters of welding materials

Temperature (℃)	Density (kg/m^3)	Thermal Conductivity (W/(m·K))	Equivalent specific heat capacity (J/(kg·K))	Elastic Modulus (GPa)	Poisson's ratio	Linear expansion coefficient ($10^{-6}K^{-1}$)	Yield Strength (MPa)
20		50.0	460	205	0.28	11.00	220
100		49.0	468	194	0.28	11.40	202
200		47.7	476	183	0.29	11.94	184
400		42.8	550	165	0.30	13.22	148
600		34.8	688	118	0.32	14.26	94
700	7820	29.6	846	86	0.34	14.62	58
750		27.0	1180	70	0.35	14.80	40
800		27.6	814	60	0.36	14.62	37
1000		30.0	720	20	0.40	13.40	25
1200		32.0	666	19.6	0.42	13.36	15
1600		330.0	1860	19	0.45	13.30	2
2000		330.0	1860	18	0.48	13.20	1

2.4. Multi-pass welding residual stress calculation scheme

The element life-and-death technique is used to simulate the sequential generation of weld bead elements and the movement of heat source. In the case that there is no highly nonlinear interaction, the indirect coupling method is more effective and convenient, because the two fields can be analyzed independently. Coupling is a cyclic process in which iterations are performed between two physics fields until the results converge to the required precision. The SOLID70 thermal analysis unit is adopted for welding temperature field analysis, which is then converted into structural analysis, and the results of temperature field are read step by step as thermal load to solve the stress field. The welding stress field analysis uses the SOLID185 structural analysis unit.

3. Design and simulation analysis of welding sequence

Numerical simulation of stress and deformation of four welding sequences of media box, the welding sequence is shown in Table 2. In the table, 1 to 10 indicate 10 weld beads from top to bottom of the alloy plate.

Table 2. Welding sequence table of dielectric box alloy plates

The scheme	The welding sequence
1	1→10→5→2→9→6→3→8→4→7
2	1→6→2→7→3→8→4→9→5→10
3	1→2→3→4→5→6→7→8→9→10
4	1→10→2→9→3→8→4→7→5→6

ANSYS was used to simulate and analyze four welding schemes of alloy plate.In order to fully analyze the influence of different welding sequences on the welding deformation of alloy plate, the transverse deformation and longitudinal deformation of alloy plate under the four welding sequences were compared respectively. The simulation results are shown in Figure 3.

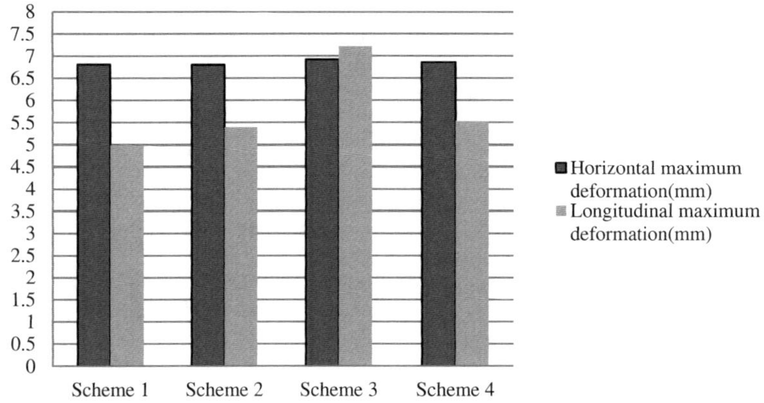

Figure 3. Maximum welding deformation of different welding sequences of alloy plates

The maximum transversedeformation of schemes 1 to 4 is 6.81mm, 6.80mm, 6.92mm and 6.86mm, respectively. It can be concluded that the maximum transversedeformation of scheme 3 is the largest, while the maximum transversedeformation of scheme 2 is the smallest, but the difference is small.Therefore, it can be concluded that different welding sequences have less influence on the transversedeformation of the alloy plate.The maximum longitudinal deformation of schemes 1 to 4 is 6.81mm, 6.80mm, 6.92mm and 6.86mm, respectively. It can be concluded that the maximum longitudinal deformation of scheme 3 is the largest, while the maximum longitudinal deformation of scheme 1 is the smallest.Therefore, it can be concluded that different welding sequences have a greater influence on the longitudinal deformation of the alloy plate.

4. Welding experiment and analysis

In order to test the simulation results of the four welding sequences of the alloy plates, 6 pieces of alloy plates of the same specification are welded in each welding sequence, and the maximum transverse deformation and maximum longitudinal deformation of alloy plates were measured. The measurement results are shown in Table 3.

The experimental results show that the average values of the maximum transverse deformation of the schemes 1 to 4 are 6.93 mm, 6.82 mm, 6.98 mm, 6.90 mm, and the average values of the maximum longitudinal deformation are 5.08 mm, 5.54 mm, 7.40. Mm, 5.57 mm.Compared with the ANSYS simulation results, the actual welding deformation is slightly larger than the simulation results, but the difference is small. The solution with the smallest welding deformation is also scheme 1, which conforms to the simulation results.

Table 3. Alloy plate welding experiment results

Schemes	Measurement type	Deformation amount(mm)						Mean values
Scheme1	transverse deformation	7.03	6.78	6.89	6.81	6.98	7.08	6.93
	longitudinal deformation	5.03	5.06	5.10	5.07	5.12	5.12	5.08
Scheme2	transverse deformation	6.81	6.80	6.79	6.89	6.84	6.81	6.82
	longitudinal deformation	5.48	5.51	5.60	5.54	5.56	5.54	5.54
Scheme3	transverse deformation	7.01	6.91	6.94	7.06	6.88	7.05	6.98
	longitudinal deformation	7.35	7.34	7.41	7.42	7.45	7.43	7.40
Scheme4	transverse deformation	6.88	6.84	6.85	6.94	6.93	6.95	6.90
	longitudinal deformation	5.63	5.52	5.54	5.57	5.60	5.53	5.57

5. Conclusion

The numerical simulation technology can predict the welding deformation during the welding process, and provide a theoretical basis for the analysis of multi-pass welding deformation of the alloy plate.The experimental results prove the correctness of the simulation results, which is of great significance for the selection of the optimal welding sequence for multi-pass welding.

Acknowledgment

This work is supported by the national natural science foundation project (the project number is 51875250).

References

[1] MurakawaHidekazu, Deng Dean, Ma Ninshu, etal. Applications of inherent strain and interface element to simulation of welding deformation in thin plate structures [J]. *Computational Materials Science*, 2012, 51(1):43-52.

[2] MitraAbhishek, PrasadN Siva, Janaki Ram GD, etal. Estimation of residual stresses in an 800 mm thick steel submerged arc weldment [J]. *Journal of Materials Processing Technology*, 2016, 229:181-190.

[3] DongWenchao, LuShanping, LiDianzhong[J].*Transactions of The China Welding Society*, 2015, 36(7).43-46.

[4] Stevenson Ryan, Toumpis Athanasios, Galloway Alexander, etal. Defect tolerance of friction stir welds in DH36 steel[J]. *Materials and Design*, 2015, 87:701-711.

[5] Park Jeong-Ung, An Gyubaek, Woo Wan chuck, etal. Residual stress measurement in an extra thick multi-pass weld using initial stress integrated inherent strain method[J]. *Marine Structures*, 2014, 39: 424-437.

IOP Conf. Series: Materials Science and Engineering **626** (2019) 012018 doi:10.1088/1757-899X/626/1/012018

Research on Quasi - Stop Control of AGV

Sun Xuan [a*], Wang Guangqi [b], Li Yingjun and Ai Changsheng

School of Mechanical Engineering, University of Jinan, Jinan, China

[a]me_sunx@ujn.edu.cn, [b]1808053595@qq.com

Abstract. Automatic guided vehicle (AGV) is one of the most important transportation tools in modern manufacturing system.In order to deliver goods to destination safely, accurately and in real time, AGV needs to have accurate parking functions. This paper first analyzes the three factors affecting AGV quasi-stop, such as load, quasi-stop initial speed and quasi-stop control trigger.Secondly, We studied the quasi - stop control system, and a quasi stop control method based on the influence factors of quasi stop is proposed. The efficiency and precision which was achieved by the method are verified experimentally. The results show that the control method improves the accuracy and efficiency of the quasi stop of AGV.

1. Introduction

Automatic guiding vehicle is a kind of transportation equipment which was widely used in modern factories to transport materials or tools from one place to another quickly, accurately and reliably[1].Therefore, in order to complete the connection with the next link and realize the normal loading and unloading work. AGV must have the function of accurate parking. Existing AGV is generally through the constant deceleration curve to slow down[2]. But, in practical work AGV will be affected by load, must stop velocity, must stop control trigger and a series of factors, this leads to use the same deceleration curve will appear larger parking error. Therefore, the study of quasi-stop control under multiple factors is of great significance.

2. QUASI - STOP INFLUENCING FACTORS

The influence factors of the magnetic navigation AGV quasi-stop control studied in this paper are mainly as follows: load, initial velocity of quasi-stop, and trigger of quasi-stop control. In practice, due to different working conditions AGV is required to carry different loads. If AGV decelerates according to the constant deceleration curve, it will cause deviation between AGV and the ideal parking point due to different inertia, resulting in a large parking error. Quasi-stopping initial velocity is also a very important factor. When the constant deceleration curve is used for deceleration and parking, too high initial quasi-stopping speed will result in AGV stop after the ideal parking position and too low initial quasi-stopping speed will result in AGV stop before the ideal parking position.

3. THE MEASUREMENT OF QUASI-STOPPING INFLUENCING FACTORS

3.1. Accurate measurement of load

If the load size is directly measured by weighing instrument, it is necessary to install weighing device on the AGV system, which will greatly increase the cost of AGV, and weighing is not very convenient.Since there is a positive correlation between the load and the current of the AGV drive motor. therefore, the load can be indirectly reflected by measuring the current of the drive motor and finding out the relationship between the load and the current.The accuracy of current measurement affects the accuracy of load. In order to realize the accurate measurement of current, the following methods are adopted:

The automatic guiding trolley studied is driven by brushless dc servo motors, and its internal connection types are connected by three-phase star bridge connection[3]. In order to realize accurate sampling of the current, hall current sensor is used to measure the current.Hall current sensor can directly connect the detected current signal with the A/D channel of the controller, and store the digital signal after A/D conversion in the controller.Figure 1 shows the principle of the internal winding of the servo motor and the installation position of the hall current sensor. It can be seen from figure 1 that phase current sampling can be realized as long as there is two-phase conduction.For example,when Q1 and Q2 are switched on, We'll know the phase current between A and B.But when the power switch transistor was shut off,the power switch transistor and a fly-wheel diode bridge arm form fly-wheel circuit, such as Q2 disconnected Q1 and D2 form a loop, the current will not flow through the hall current sensor, the hall current sensor can't detect the current signal so current sensor can only be detected during the power transistor opening current signal[4].

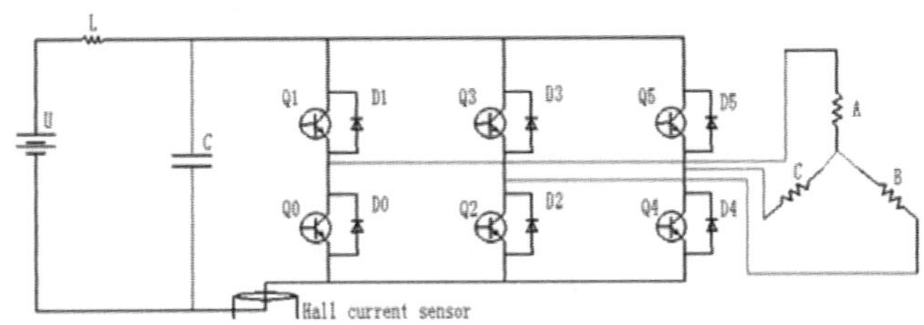

Figure 1. Schematic diagram of installation position of motor current sampling sensor.

The power transistor generates peak current at the moment of opening and closing, which can' t accurately express the relationship with the load[5]. Therefore, it is necessary to design appropriate sampling points to avoid peak current.To ensure that in each switch cycle to determine a fixed sampling points, and away from the switch point, using the controller in the timer trigger A/D conversion and guarantee sampling during the midpoint of the power transistor opening period the power transistor opening (during high level of PWM) the sampling method of average current so well to avoid the effects of the switch noise, guarantee the accuracy of sampling[6].The sequence diagram of sampling control is shown in figure 2.

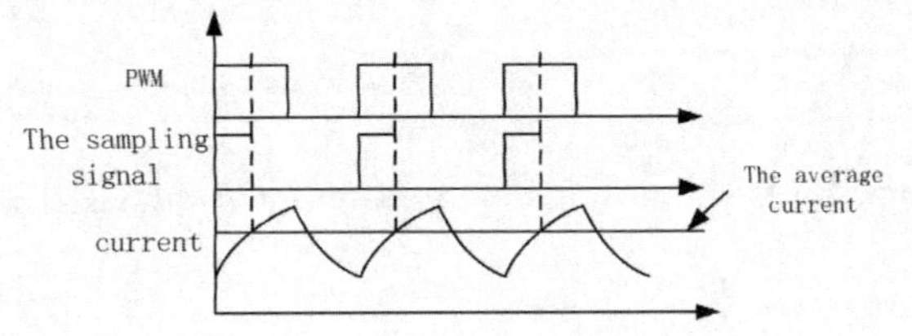

Figure 2. Current sampling control timing diagram.

3.2. Quasi-stop initial velocity measurement

Study adopts brushless dc servo motor drive with velocity feedback port, synchronization with the operation of the motor output shaft, the motor output shaft turn each output a certain number of pulse signal, so that the rotational speed of the motor can be calculated by measuring the output pulse frequency of output port .

3.3. Quasi - stop control trigger

In order to realize accurate control of accurate trigger, this research USES laser ranging to trigger, as shown in figure 3 ,the AGV vehicle equipped with a laser range sensor, when the RFID card reader on the AGV detected deceleration of the L2 distance reflection plate location site, trigger laser range sensor, laser ranging sensor launch a laser beam to the target reflection board, according to the laser beam to be accepted by the launch time, with the target via internal algorithm calculates the AGV baffle distance, when detected the distance of AGV and baffle is L1 triggered slowing down, to realize the quasi - stop control of accurate trigger.

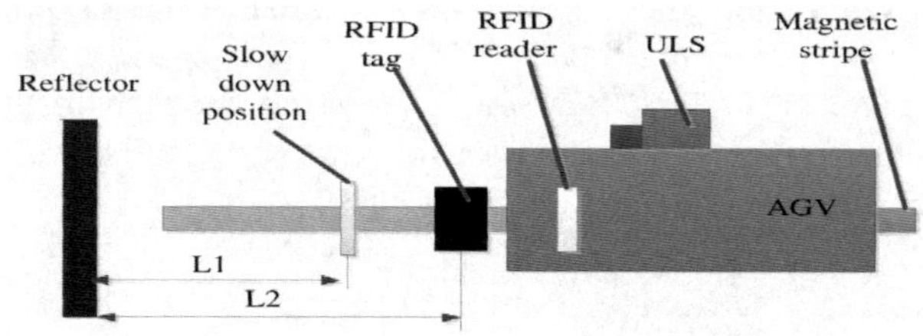

Figure 3. AGV quasi - stop control trigger schematic diagram.

4. THE COMPOSITION OF QUASI - STOP CONTROL SYSTEM

The quasi-stop control system of AGV is mainly composed of the controller execution module, information acquisition module, human-computer interaction module and safety module. As shown in figure 4.

Figure 5 shown the control process of the control system, information collection module sends the information to the controller, the controller send control commands to perform after data processing module, execution module to perform the relevant orders, at the same time, the controller will collect data information sent to the human-computer interaction module, convenient for offline data processing, when security module found obstacles,The obstacle information is sent to the controller, and after the information processing, the controller sends instructions to the driver module, and the driver module carries out relevant actions.

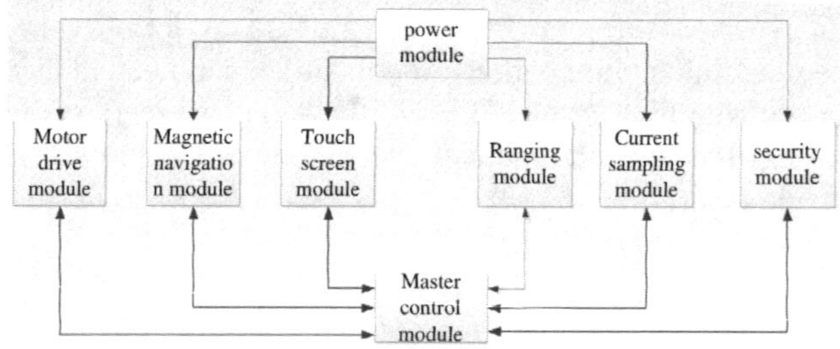

Figure 4. AGV quasi - stop control system block diagram.

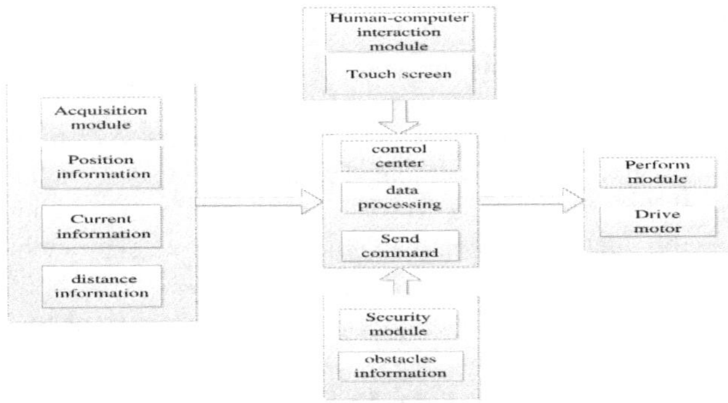

Figure 5. Workflow of quasi - stop control.

5. Establishment method of quasi-stop control deceleration model

In order to establish the deceleration model, PI closed-loop speed control is adopted, and the establishment process is as follows: As shown in figure 6 when AGV run under a load to the deceleration point D, according to the experience setting goal speed points between D and E in some position, through the speed sensor to the system of real-time feedback current speed, the laser ranging sensor to the system real-time feedback distance information and by the analysis of controller of distance and velocity, corresponding PI control parameters are given,to ensure that the target speed is reached at the specified position and the control block diagram is shown in figure 6.When the AGV is parked observe the error between the parking position and the target position.according to the error size and vehicle running status, change the PI control parameters.continue to experiment,observing the parking point error and the stability of vehicle running.and continue to modify the PI control parameters, Until the desired parking position and operating status are reached, record speed curve at this time.Changing the load of AGV and continuing the experiment.After several sets of experiments, the ideal speed curves under different loads as shown in Fig.7. the speed curves were analyzed and processed. Fitting the deceleration model between the load and the speed curve.Experiment and observation, to determine the model of quasi-stop control deceleration.

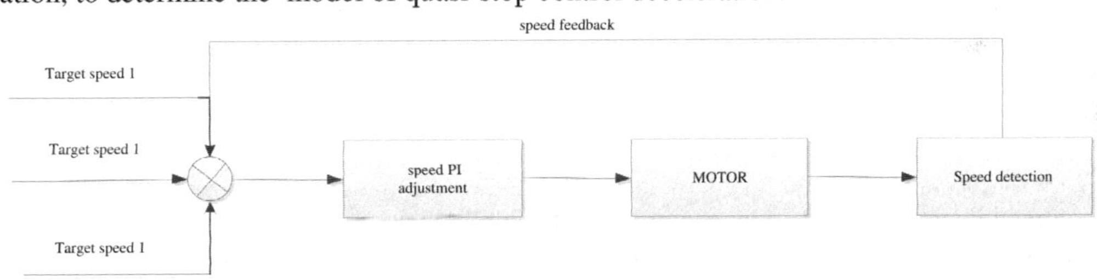

Figure 6. PI control block diagram.

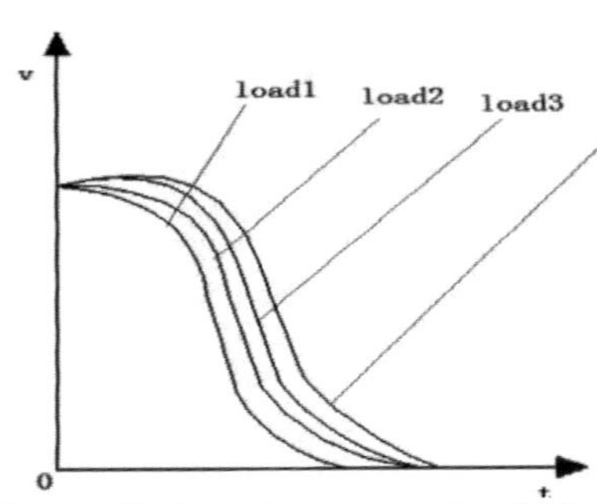

Figure 7. Speed graph under different loads.

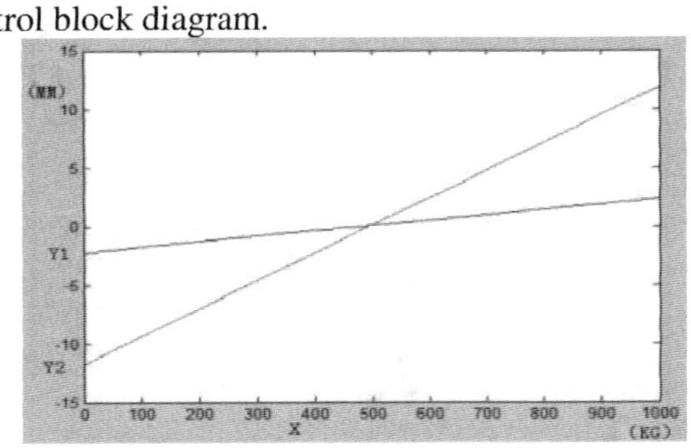

Figure 8. AGV accurate parking accuracy.

6. EXPERIMENTAL VERIFICATION OF QUASI - STOP CONTROLEFFECT

To illustrate the effectiveness of the must stop control method and accuracy, the same velocity, the same stopping distance under different load, comparing the control error according to the constant deceleration curve and using quasi control method reduction of parking to slow down and stop. As shown in figure 8, Y2 is according to the constant deceleration curve to reduction and after then the relationship between load and parking error curve. Y1 is according to the constant deceleration curve to reduction and after then the relationship between load and parking error curve.

It can be seen from figure 8 that the deceleration and parking error in accordance with the constant deceleration curve is approximately within 10mm. After deceleration in accordance with the quasi-stop control method, the parking error can be controlled within 2mm, and the parking accuracy has been greatly improved.

7. Conclusion

In order to improve the parking precision of the AGV, a must stop control method is discussed in this paper, the method solves issue that the AGV constant deceleration curve under different load have large stop error , improves the efficiency of the quasi stop precision of the AGV, at last experiments show that the control method can reduce the error from 10mm to within 2mm, which indicates the feasibility and effectiveness of the control method and provides a feasible method for the accurate parking of AGV.

Acknowledgment

This work is supported by the national natural science foundation project (the project number is 51875250).

References

[1] Iris F. A. Vis, Survey of Research in the Design and Control of Automated Guided Vehicle Systems, *Ruropean 10 urnal of Operational Research*, Volume 170, Issue 3, I May 2012, pp. 677-709.

[2] Lothar Schulze, Sebastian Behling, and Stefan Buhrs, Automated Guided Vehicle Systems: a Driver for Increased Business Performance, *Proceedings of the International Multi Conference of Engineers and Computer Scientists* 2012, Volume II, IMECS 2012, 19-21 March, Hong Kong, 2012.

[3] Bob Trebilcock.Automatic Guided Vehicle (AGV)Basics[J].*Material Handling Product News*,2016 (12) 65-68.

[4] Lazaro J L, Garcia J C, Mazo M, Disibuted architecture for control and path planning of autonomous vehicle[J]. *Microprocessors and Microsystems*, 2015, 25: 159-166 .

[5] Rogers,Lorie King.Automatic Guided Vehicles[J]. *Modern Materials Handing*, 2013, 66(9), 43-46.

[6] Trebilcock, Bob.Automatic guided vehicles streamline operations[J]. *Modern Materials Handling*, 2014, 65(10): 18-23.

IOP Conf. Series: Materials Science and Engineering **626** (2019) 012019 doi:10.1088/1757-899X/626/1/012019

Research on Plant Control Robotic Wire Control Chassis Control System

Yongcheng Du, Changsheng Ai*, Xuan Sun, YequanBao, Weixin Li, Hui Zhang

School of Mechanical Engineering, University of Jinan, Shandong Province, China

me_aics@ujn.edu.cn

Abstract. In order to facilitate the independent operation of the wheeled plant protection robot, a wire-controlled chassis with wire control, modularization and generalization was designed and manufactured. The steering drive mode of the plant protection robot adopts front wheel drive rear wheel steering .The steer-by-wire system uses a steering motor to drive the steering gear and installs a corner sensor on the steering column to achieve closed-loop control of the steering. The throttle-by-wire system uses a PWM speed control method and uses a Hall sensor to achieve closed-loop speed control. The brake-by-wire system uses a dedicated drive to control the electric linear actuator for braking and closed loop control of the brake through the current loop. For the communication method, the CAN network is used to build the communication of each module. The results show that the line-controlled site can accurately follow the expected action value given by the decision-making system to ensure the accuracy and stability of the plant protection robot operation.

1. Introduction

The level of automation of plant protection machinery in China is low. Most of them are backpack and traction operation, which requires manpower to assist the operation. The efficiency is low and the waste is serious[1]. Therefore, to improve the automation level of domestic plant protection machinery and liberate people from the plant protection work with high labor intensity and bad working environment has become the most important thing to realize the modernization of agriculture in China. However, as the end-effector, the wire-controlled chassis is the basis to ensure the effective operation of plant protection robot. Therefore, in order to improve the automation level of plant protection machinery in China, it is particularly important to develop and design a line-controlled chassis which integrates steering, walking and braking[2].

The prototype of wire control technology came into being in the 1950s[3]. TRW in the United States and Kasselmann in Germany replaced the mechanical structure between the steering wheel and the steering wheel with corresponding control signals.In 2017, Tesla put forward a more innovative scheme for the layout of the wire control system, which simulates human driving behavior and achieves driverless driving from Los Angeles to New York. Domestic enterprises started online control technology relatively late, and the research of online control technology in major universities mainly stayed at the theoretical stage.On the basis of on-line steering control of Baidu UAV in 2017, combined with Road Hacker platform, the test of hybrid automatic driving of loop road and highway road conditions was realized for the first time in China[4]. The wire-controlled chassis is mainly composed of three systems: steering, throttle and brake. Among them, there are two main forms of steer-by-wire: 1 cancel the mechanical connection between the steering wheel and the steering actuator, increase the redundancy of the system through multiple motors and controllers; 2 add an electromagnetic clutch between steering wheel and steering actuator as failure backup to increase

IOP Conf. Series: Materials Science and Engineering **626** (2019) 012019 doi:10.1088/1757-899X/626/1/012019

system redundancy. The main suppliers at home and abroad are Bosch, TRW and Zhejiang Wanda. The throttle-by-wire has been developed so that the ECU can determine the opening and closing of the throttle, the amount of fuel injection, and the injection time interval based on the position of the accelerator pedal. The main suppliers abroad are Bosch, Continental and so on. Brake-by-wire has been developed into two types of EHB(Electronic Hydraulic Brake) and EMB(Electromechanical Brake System). Among them, EHB is more mature, with the advantages of small size, light weight, fast response speed and more flexible operation. The main suppliers at home and abroad are Bosch, Lianchang Electronics and WanxiangGroup[5].

Considering the high cost of wire-controlled chassis, the fact that suppliers do not retail and control protocols are not open to the outside world, it is necessary to design a low-speed automatic driving wire-controlled chassis for the autonomous operation experiment of plant protection robots.

2. TRANSMISSION SCHEME AND STEERING DRIVE SCHEME FOR REMOTE CONTROL CHASSIS

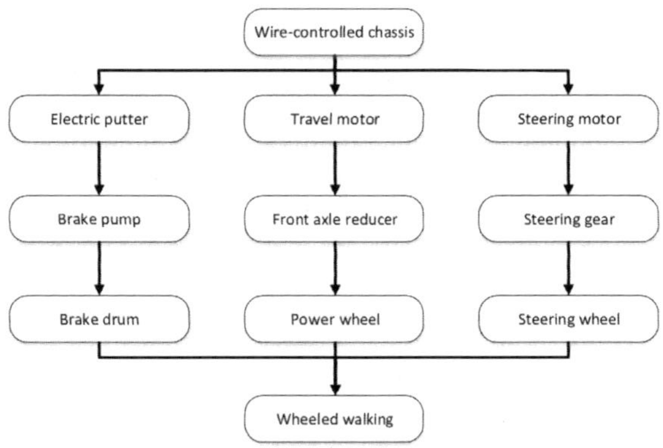

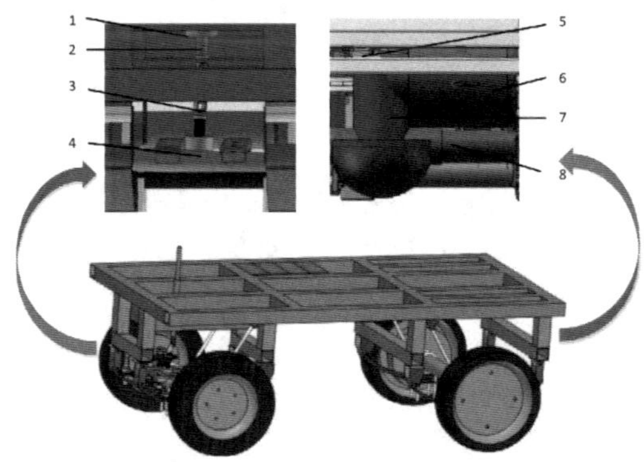

Figure 1. Transmission system block diagram**Figure 2.** Model of the line control chassis

The steering drive mode of the plant protection robot adopts front wheel drive rear wheel steering .As shown in Figure 1, the transmission system of the wire-controlled chassis adopts the traveling motor and the reducer to drive to realize the chassis walking. The electric linear actuator is used to push the hydraulic Brake to realize the braking of the wire-controlled chassis, and the steering motor is used with the steering machine. The form is used to achieve the steering of the wire-controlled chassis.

As shown in Figure. 2, the steering part comprises a 1 angle sensor, 2 steering motor, 3 coupling, 4 Steering gear; the energy device is 5 battery pack; the throttle part comprises 6 walking motor, 7 speed reducer and 8 transmission front axle.

3. WIRE-CONTROLLED CHASSIS CONTROL SYSTEM

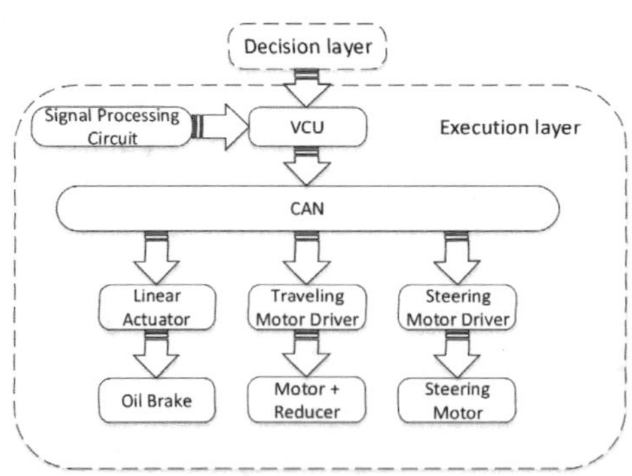

Figure 3. Control schematic of the wire control chassis

IOP Conf. Series: Materials Science and Engineering **626** (2019) 012019 doi:10.1088/1757-899X/626/1/012019

The battery pack is placed in the middle of the chassis structure to facilitate the supply of power to each electronic module.Figure3 shows the decision layer outputs the expected value of the next action to the execution layer. The VCU in the line control site will perform linear control of the brake, throttle and steering according to the input information of the sensor. The steering is realized by the VCU sending a command to the drive of the steering motor, the drive drives the steering motor to rotate, and the steering motor is connected to the steering gear through the universal joint, thereby transmitting the torque of the steering motor to the steering gear through the coupling. To push the rear wheel for steering[6]. The realization of the walking is also to send a command to the drive of the traveling motor through the VCU, thereby controlling the traveling motor to rotate, the motor reducer is placed at the front center, and the power is applied to the driving wheel through the front axle. When the speed of the plant protection robot is too large and requires braking, the ECU sends a command to the electric linear actuator, and the electric linear actuator action changes the flow direction of the hydraulic oil in the hydraulic circuit, thereby controlling the hydraulic cylinder action to push the brake pad to clamp the driving wheel to generate the brake. effect.

3.1. Steering control system for wire control chassis

As shown in the figure 4: the upper computer gives the VCU a command to rotate the target angle α_{aim}, and the VCU controls the steering motor of the steering wheel by the drive of the steering motor. At the same time, the steering angle sensor is mounted on the steering column to feed back the actual rotation angle α and the target. The transfer α_{aim} comparison yields a deviation $\Delta\alpha$, which is input to the VCU, and the controller will adjust the corresponding rotation angle according to the positive and negative deviations of the deviation. When the deviation is positive, the steering motor rotates forward, and vice versa; when the deviation value is large, the controller outputs a PWM (pulse width modulation) square wave with a large duty ratio[7]. At the same time, theangle sensor will also monitor the corner size in real time and compare it as the input amount and the devi the set threshold, the current steering control will also end. The VCU is the control system, the steering mechanism is theactuator, the steering motor is the controlled object, and the transfer sensor is used as the feedback measuring device. These four parts constitute the closed-loop servo steering system.

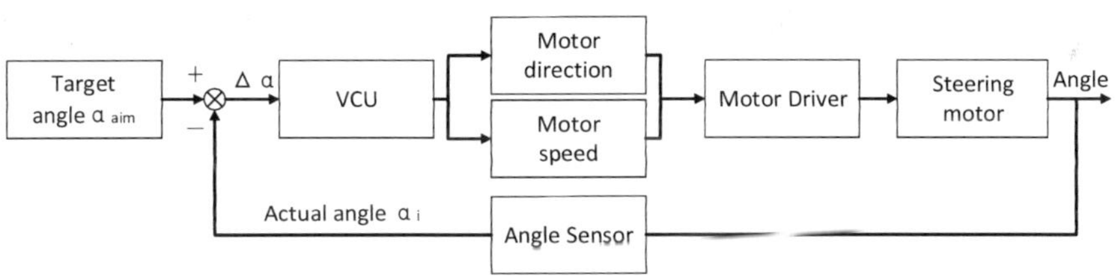

Figure 4. Steering control schematic

Based on the above principle, it is proposed to use DC servo motor as the steering motor. The interface description of the corresponding driver is shown in Figure 5. The CAN bus communication method makes the vehicle wiring simpler, more efficient and more reliable. Each serial port command sent by the VCU to the drive is composed of 10 bytes. The mode selection of the drive is shown in Figure 6. Here we plan to select the mode of the speed position and write the corresponding address against the corresponding communication protocol. The control flow is as follows:

- Send reset command.
- Waiting for 500ms.
- Send mode select command to put the drive into a mode.
- Wait500ms
- Send a data command in the mode that has been entered (This instruction is sent periodically, with a minimum interval of 1ms and a recommended interval of 10ms).

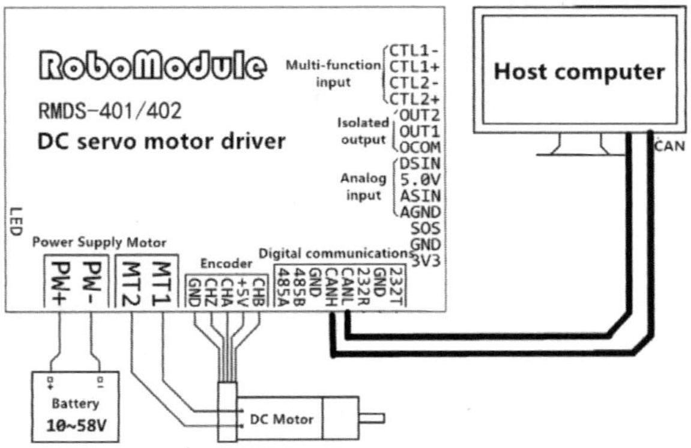

Figure 5. Steering motor drive interface description

The commands that the driver feeds back to the VCU mainly include the following three types: 1. Sending current, speed, position and other information; 2. Sending CTL1 and CTL2 port level status to the outside; 3. Sending online detection feedback.

3.2. Throttle control system for wire control chassis

In order to ensure the uniformity of application of plant protection robots during farm operations, it is necessary to ensure the ability of cruise control under complex road conditions. Therefore, we use a DC48V, 3KW travel motor with speed feedback, the drive speed signal is 0-5V analog, using pulse width modulation (PWM) method, the constant DC voltage is modulated into a variable size and polarity. TheDC voltage is used as the armature voltage of the motor. Since the accuracy of the motor speed is very high, and the load of the motor changes greatly during the operation, the speed closed-loop control is adopted, and the method realized by using the DC brushless motor can measure the signal of the Hall signal line. obtain Speed and travel distance information[8].

As shown in the figure 6: when the VCU receives the target speed Vaim issued by the host computer, it sends a command to the driver to control the speed of the traveling motor, and uses the photoelectric encoder as the speed measuring component to realize real-time measurement of the motor speed and obtain the motor speed information Vi. Feedback to the VCU to form a closed loop speed control system.

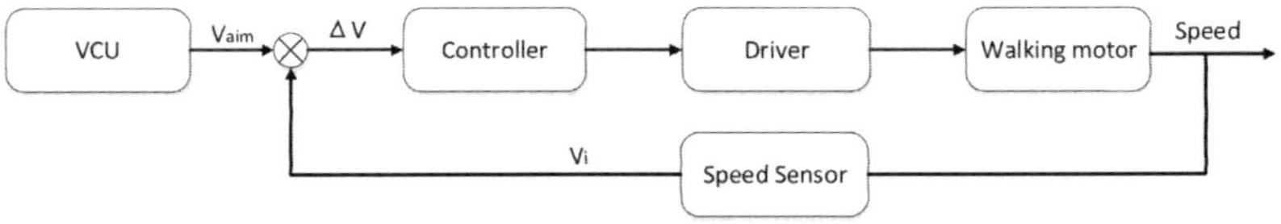

Figure 6. Throttle control schematic

3.3. Brake control system for wire control chassis

In order to ensure that the plant protection robot accurately decelerates or brakes during the operation[9], for the brake system of the wire-controlled chassis, we add an electric linear actuator to replace the pedal action of the person on the basis of the traditional hydraulic Brake system, and the corresponding work The flow is shown in Figure 7. The electric linear actuator action pushes the hydraulic oil in the cylinder into the brake caliper along the hydraulic line, and then tightens the hub to achieve braking.

In order to realize the closed-loop control of the line control, the driver of the model AQMD3610NS-A is used, and the braking torque and the current are used to realize the feedback of the braking torque through the current feedback. The method has the advantages of fast response speed, light weight, no complicated hydraulic system and convenient maintenance, and the corresponding schematic diagram is shown in Figure 8 .

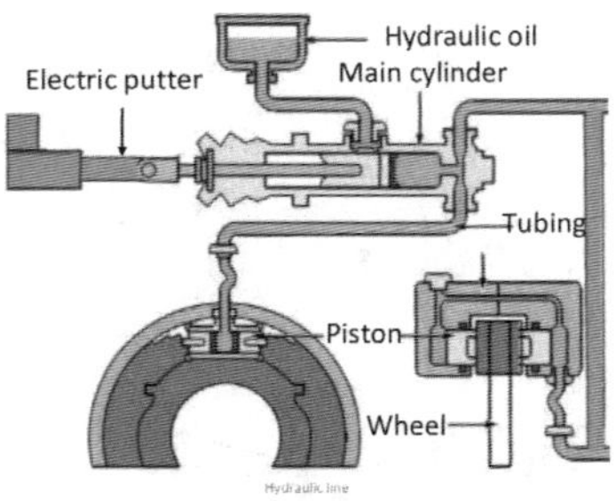

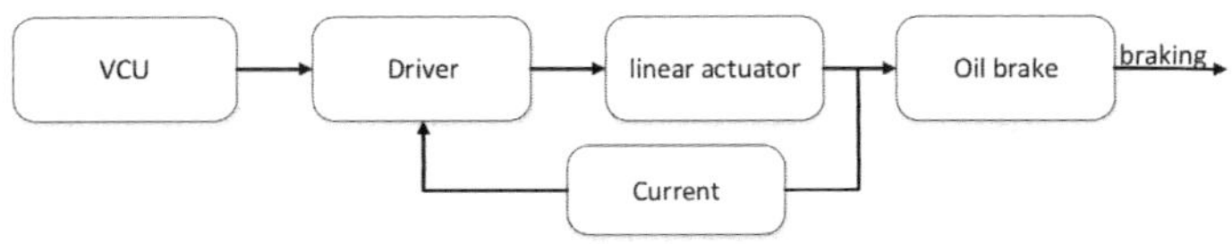

Figure 7. Line control dynamic schematic **Figure 8.**Brake control schematic

4. CONCLUSION

Taking the self-operating plant protection robot as the research object, the transmission mechanism, steering drive mode, steering, walking and braking of the wire-controlled chassis are designed and manufactured, and the closed loop control of the position loop, speed loop and current loop are applied respectively. The test results show that the line control site can accurately execute the output command of the decision-making layer, realize the stable closed-loop control of the traveling motor and the steering motor, and the fast response control of the braking system. It has the characteristics of modularization and intelligence to ensure the operation of the plant protection robot's flexibility and stability of driving.

References

[1] Van Zanten A T, Erhardt R, Landesfeind K, et al. VDC system development and perspective. *SAE Paper 980235*, 2010.
[2] Leffler H. Consideration of lateral and longitudinal vehicle stability by function enhanced brake and stability control system. *SAE 940832*， 2012.
[3] Straub A. DSC (Dynamic Stability Control) in BMW 7 series cars. *International Symposium on Advanced Vehicle Control*, 2009.
[4] United States National Highway Traffic Safety Administration.*FMVSS No. 126 Electronic Stability Control Systems,NHTSA Final Regulatory Impact Analysis*.2007.
[5] TIAN Cheng-wei. Research on Fault Tolerant Control Method for Steer-by-wire Automobile [D].*Changchun:Jilin University*,2010.
[6] HE Lei.Research on Dual-motor Control Method Based on FlexRay Bus for Steering-by-wire Automobile[D].*ChangChun:Jilin University*,2011.
[7] Falcone P, Tseng H E, Borrelli F, et al. MPC- based yaw and lateral stabilisation via active front steering and braking. *Vehicle System Dynamics*, 2008, 46 (sup1) 611~628.
[8] DAIROU V,PRIZE A,SIEFEFMANN J,et al. An Original Method to Predict Brake Feel:A Combination of Design of Experiments and Sensory Science [J].*SAE Paper* 2013-01-0598.
[9] MetinErsoy, Stefan GiesHrsg. Automobile chassis manual[M]. *Walluf und Wolfsburg: Springer*, 2017: 430-436.

IOP Conf. Series: Materials Science and Engineering **626** (2019) 012020 doi:10.1088/1757-899X/626/1/012020

Analysis Line Approach for Assembly Tolerance Analysis

Heping Peng [1]* **and Zhuoqun Peng** [2]

1 School of Electromechanical & Architectural Engineering, Jianghan University, Wuhan 430056, China

2 School of Machinery and Automation, Wuhan University of Science and Technology, Wuhan 430081, China

hbjhun_penn@outlook.com

Abstract. This work presents the analysis line approach for assembly tolerance analysis. Firstly, the product functional requirements and geometric variations of part positioning features can be converted into small displacements of a series of analysis points along the specified analysis directions. The small displacement torsor (SDT) theory is employed to establish the mathematical relations between the displacements of these analysis points. And then, the variation transfer relations and tolerance accumulation calculations of various orthogonal joints between different parts are studied to determine the maximum displacement of the analysis points on the functional surface. The final displacement accumulation results of the functional surface are compared with the geometric functional requirements of the product, so as to judge whether the product functional requirements are met. Finally, an example is given to demonstrate the effectiveness of the proposed analysis line approach.

1. Introduction

Tolerance analysis involves identifying the related tolerances in an assembly or a mechanism and calculating the total stack-up of these related tolerances. It is a powerful tool to predict the effects of dimensional and geometric tolerances that have been allocated to the components of an assembly on the assembly functional requirement (FR) itself [1]. Many researches focus on three-dimensional tolerance design technology, especially three-dimensional tolerance analysis technology, and various tolerance analysis model have been developed, such as matrix model[2], vector loop model[3], unified Jacobian-torsor model[4], T-map model[5], etc. Anselmetti [6-7] proposed the analysis line method to perform tolerance analysis by using deviation transfer relation equations to establish the accumulation relations between each tolerance and the geometric FR. Later, Anselmetti extended this method to braze welded assemblies [8]. The purpose of this paper is to further explore the application of analysis line approach in assembly tolerance analysis, especially when the geometric tolerance principles such as: maximum material requirement (MMR), least material requirement (LMR), etc. are taken into account.

2. Analysis line approach

2.1 Definition of analysis line

Functional requirement usually specified by the designer is the variation of the dimension or angle between two geometric features, or the position or orientation tolerance of a geometric feature relative to the datum features. The surface feature of the FR specification is also called functional surface. For tolerance analysis based on the analysis line approach, it is usually necessary to discretize the functional surfaces. That is, the geometrical variation of the functional surface is represented by the displacements of finite discrete points on the surface towards the discrete directions. These discrete points are called analysis points, the discrete directions are called analysis directions, and the analysis lines are defined by these analysis points and the analysis directions. Generally speaking, as long as the displacement of the analysis point on the functional surface at which the position changes most towards the analysis direction meets the tolerance specified by FR, the product design complies with the functional requirements.

2.2 Discretization of the functional surface

Different feature types and geometrical tolerance items on a functional surface have different discretization modes. Figure 1 displays the discretization modes of several typical functional surfaces. For the plane feature with a positional tolerance requirement, the analysis point P_i is distributed on the boundary of the plane, and the analysis directions **p** and -**p** are perpendicular to the plane. In particular, for the non-circular plane, the analysis point should be located at the vertex of the plane boundary, as shown in figure 1 (a). For the plane feature with an orientation requirement (such as parallelism tolerance, perpendicularity tolerance, etc.) as shown in figure 1 (b), the analysis directions are discretized to 8 $\mathbf{p}_i$ directions. For the cylinder feature with a positional tolerance requirement, the analysis points P_1 and P_2 are the two endpoints of the cylinder. If the tolerance zone is an area between two parallel planes, the analysis direction is the vector direction **p** and -**p** perpendicular to the cylindrical axis as shown in figure 1 (c); if the tolerance zone is a cylinder area, the analysis directions are the directions indicated by 8 vectors $\mathbf{p}_i$ perpendicular to the cylindrical axis shown in figure 1 (d).

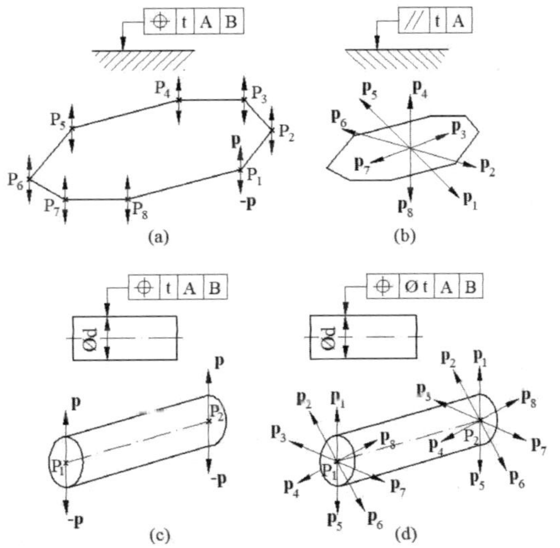

Figure 1. Discretization of ending surface.

Here, the displacement of point P along **p** direction relative to its nominal position in *R* datum reference frame is denoted as $\varepsilon^R(P, \mathbf{p})$; the rotation angle of the analyzed geometric feature E (plane, axis, etc.) around **p** direction relative to its nominal position in *R* datum reference frame is denoted as $\varphi^R(E, \mathbf{p})$.

3. Three-dimensional variations transfer of orthogonal joints

In order to make a part have a definite position during assembling process, we adopt the six-point positioning rule to fix the part and these six positioning points are respectively located on three mutually perpendicular planes, as shown in figure 2.The main positioning plane is defined by three points M_1, M_2, and M_3; the 2nd positioning plane is defined by points S_1 and S_2; the 3rd positioning plane is defined by one point T. The normal unit vectors of these three positioning surfaces are denoted as **m**, **s** and **t**, respectively. In this case, the displacement of the analysis point P on the functional surface in the analysis direction **p** is written as [7]:

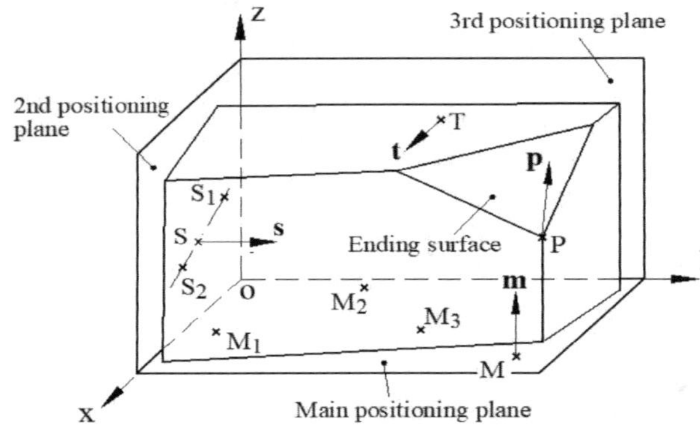

Figure 2. Positioning on three perpendicular planes.

$$\varepsilon(\mathrm{P},\mathbf{p}) = k_m \cdot \varepsilon(\mathrm{M},\mathbf{m}) + k_s \cdot \varepsilon(\mathrm{S},\mathbf{s}) + k_t \cdot \varepsilon(\mathrm{T},\mathbf{t}) + \varphi \cdot [\mathbf{OP}\times\mathbf{p} - k_m \cdot (\mathbf{OM}\times\mathbf{m}) - k_s \cdot (\mathbf{OS}\times\mathbf{s}) - k_t \cdot (\mathbf{OT}\times\mathbf{t})] \quad (1)$$

where the influence coefficients $k_m=p_z/m_z$, $k_s=p_y/s_y$, $k_t=p_x/t_x$.

Equation (1) establishes the transfer relation between the displacement of the point on the functional surface of the part and the displacements of the feature points on these three positioning surfaces of this part. The influence coefficient shows the contribution rate of the each positioning surface variation to the displacement of the functional surface. Obviously, when the variation of the positioning surface increases, the displacement of the functional surface will also increase, and vice versa, so these three influence coefficients must be all positive. Accordingly, we can determine the directions of vectors **m**, **s** and **t**. According to equation (1), in order to find the points M and S whatever φ rotation, we let:

$$\mathbf{OP}\times\mathbf{p} - k_m \cdot (\mathbf{OM}\times\mathbf{m}) - k_s \cdot (\mathbf{OS}\times\mathbf{s}) - k_t \cdot (\mathbf{OT}\times\mathbf{t}) = 0 \quad (2)$$

According to equation (2), we can get the coordinates of points M and S.

$$\begin{cases} x_m = x_p - p_x(z_p - z_t)/p_z \\ y_m = y_p - p_y(z_p - z_s)/p_z \\ z_m = 0 \end{cases} \quad (3)$$

$$\begin{cases} x_s = x_p - p_x(y_p - y_t)/p_y \\ y_s = y_1 + (y_2 - y_1)(x_s - x_1)/(x_2 - x_1) \\ z_s = z_1 + (z_2 - z_1)(x_s - x_1)/(x_2 - x_1) \end{cases} \quad (4)$$

where point S is on the straight line defined by point $S_1(x_1, y_1, z_1)$ and point $S_2(x_2, y_2, z_2)$.

Thus, the positions of these three points M, S, and T can be determined. Theses points can also be called analysis points. The analysis point M and vector **m** define an analysis line which is called the main analysis line. Similarly, the analysis point S and vector **s** define the second analysis line; the analysis point T and vector **t** define the third analysis line. Then, as long as the maximum displacements at these three points are obtained, the maximum displacement of the analysis point P on the functional surface can be calculated.

4. Variations transfer of the cylinder link

4.1 Displacement of analysis point on cylinder with MMR

For the cylinder shaft-hole fit with MMR as shown in figure 3, the maximum material size of the cylinder shaft is: $d_M = d + t_d$; The maximum material virtual size of the cylinder hole defined by the positional tolerance t_{po} is: $D_M = D-t_D-t_{po}$; The maximum material virtual size of the cylinder hole defined by the parallelism tolerance t_{pa} is: $D_O = D-t_D-t_{pa}$. The maximum displacements of points M_1 and M_2 are respectively:

$$\varepsilon\ (M_1, \mathbf{m}) = (D_M-d_M)/2 \quad (5)$$

$$\varepsilon\ (M_2, \mathbf{m}) = (D_M-d_M)/2 + (D_O-d_M)\cdot l_2/l_1 \quad (6)$$

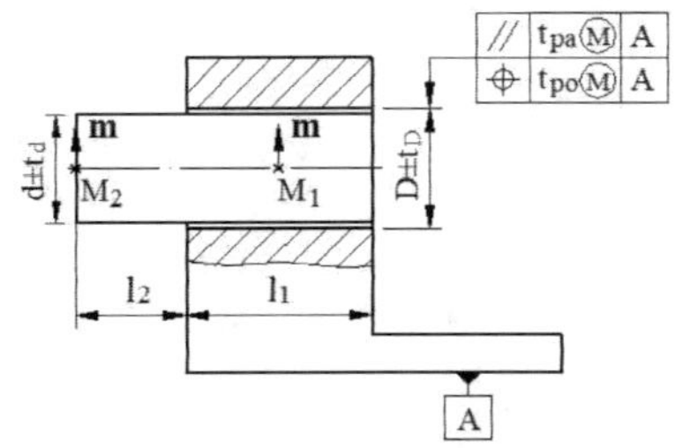

Figure 3. The displacement of cylinder link with MMR.

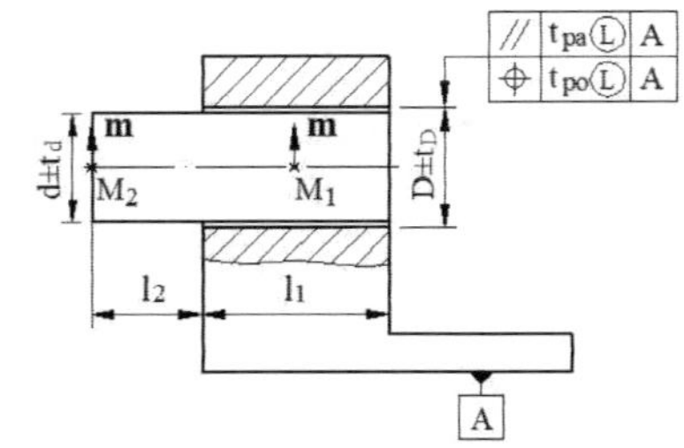

Figure 4. The displacement of cylinder link with LMR.

4.2 Displacement of analysis point on cylinder with LMR

For the cylinder shaft-hole fit with LMR as shown in figure 4, the displacement of the point on the cylinder shaft can be regarded as caused by the clearance, and this displacement has a maximum value when the clearance reaches its maximum. The least material size of the cylinder shaft is: $d_L = d-t_d$. The least material virtual size of the cylinder hole defined by the positional tolerance t_{po} is: $D_L = D+t_D+t_{po}$. The least material virtual size of the cylinder hole defined by the parallelism tolerance t_{pa} is: $D_O = D+t_D+t_{pa}$. The maximum displacements of points M_1 and M_2 are respectively:

$$\varepsilon\ (M_1, \mathbf{m}) = (D_L-d_L)/2 \quad (7)$$

$$\varepsilon\ (M_2, \mathbf{m}) = (D_L-d_L)/2 + (D_O-d_L)\cdot l_2/l_1 \quad (8)$$

4.3 Variations transfer of the cylinder-plane joint

The joint types between parts in 3-D assemblies mainly include plane-plane, plane-cylinder, cylinder-cylinder, etc, and most of which are orthogonal joins. In this case, the established datum system is usually an orthogonal coordinate system. Next we further discuss the variation transfer of cylinder-plane joint.

As shown in figure 5, the centring pin is positioned by a cylindrical surface and a plane perpendicular to the axis of the cylindrical surface, the cylindrical surface and the plane are the main positioning surface and the second positioning surface, respectively. The positioning datum system is established by taking the cylinder axis as y-axis and the line perpendicular to the cylinder axis as z-axis. The main analysis point M $(0, y_m, 0)$ is on the cylinder axis with the vector direction **m** $(0, 0, m_z)$; the second analysis point S $(0, y_s, z_s)$ is on the second positioning plane with the vector direction **s** $(0, s_y, 0)$.

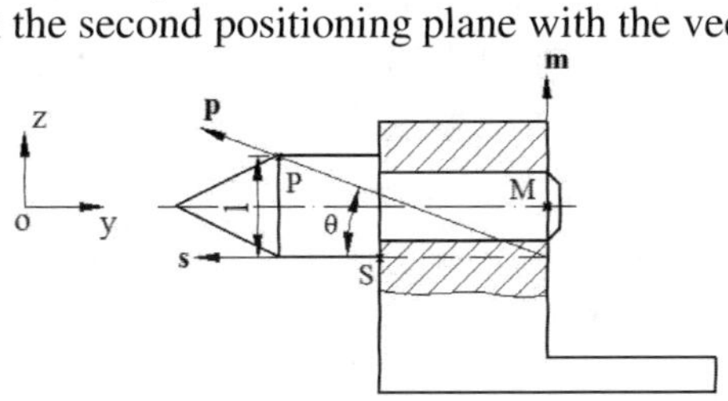

Figure 5. Cylinder-plane joint

The displacement of the analysis point **P** $(0, y_p, z_p)$ on the functional surface along the vector direction **p** $(0, p_y, p_z)$ will be:

$$\varepsilon(\mathrm{P},\mathbf{p}) = \mathrm{v} \cdot \mathrm{p}_y + \mathrm{w} \cdot \mathrm{p}_z + \varphi \cdot (\mathbf{OP} \times \mathbf{p}) \tag{9}$$

The displacement of the analysis M (0, y_m, 0) along the **m** (0, 0, m_z) will be:

$$\varepsilon(\mathrm{M},\mathbf{m}) = \mathrm{w} \cdot \mathrm{m}_z + \varphi \cdot (\mathbf{OM} \times \mathbf{m}) \tag{10}$$

The displacement of the analysis S (0, y_s, z_s) along the **s** (0, s_y, 0) will be:

$$\varepsilon(\mathrm{S},\mathbf{s}) = \mathrm{v} \cdot \mathrm{s}_y + \varphi \cdot (\mathbf{OS} \times \mathbf{s}) \tag{11}$$

Eliminating v, w in the above equations (9)-(11), one will get:

$$\varepsilon(\mathrm{P},\mathbf{p}) = \mathrm{k}_m \cdot \varepsilon(\mathrm{M},\mathbf{m}) + \mathrm{k}_s \cdot \varepsilon(\mathrm{S},\mathbf{s}) + \varphi \cdot [\mathbf{OP} \times \mathbf{p} - \mathrm{k}_m \cdot (\mathbf{OM} \times \mathbf{m}) - \mathrm{k}_s \cdot (\mathbf{OS} \times \mathbf{s})] \tag{12}$$

where $k_m = p_z/m_z$, $k_s = p_y/s_y$. We can orientate the vectors **m** and **s** by setting k_m and k_s positive. Regardless of the value of φ, in order to find points M and S, the rotation component of equation (12) needs to be eliminated. So one get:

$$\mathbf{OP} \times \mathbf{p} - \mathrm{k}_m \cdot (\mathbf{OM} \times \mathbf{m}) - \mathrm{k}_s \cdot (\mathbf{OS} \times \mathbf{s}) = 0 \tag{13}$$

When $p_z \neq 0$, $z_p - z_s \neq 0$, we will have:

$$p_y/p_z = (y_p - y_m)/(z_p - z_s) \tag{14}$$

In this case, the three analysis lines intersect at one point as shown in figure 5. So one get:

$$\varepsilon(\mathrm{P},\mathbf{p}) = \mathrm{k}_m \cdot \varepsilon(\mathrm{M},\mathbf{m}) + \mathrm{k}_s \cdot \varepsilon(\mathrm{S},\mathbf{s}) = \sin\theta \cdot \varepsilon(\mathrm{M},\mathbf{m}) + \cos\theta \cdot \varepsilon(\mathrm{S},\mathbf{s}) \tag{15}$$

When $p_z=0$, we have θ=0, the rotation component φ is not equal to zero. The displacement of the analysis point **P** (0, y_p, z_p) along the vector direction **p** (0, p_y, p_z) will be:

$$\varepsilon(\mathrm{P},\mathbf{p}) = \mathrm{k}_s \cdot \varepsilon(\mathrm{S},\mathbf{s}) + \varphi \cdot [\mathbf{OP} \times \mathbf{p} - \mathrm{k}_s \cdot (\mathbf{OS} \times \mathbf{s})] = \varepsilon(\mathrm{S},\mathbf{s}) + \varphi \cdot (\mathbf{SP} \times \mathbf{p}) = \varepsilon(\mathrm{S},\mathbf{s}) + \mathrm{l} \cdot \mathrm{a}(\Delta,\mathbf{a}) \tag{16}$$

where l is the distance between P and S in the direction of the main analysis line; Δ is the axis of the main cylinder surface; a(Δ, **a**) is the maximum rotation angle of axis Δ around vector **a** (**a=SP**×**p**). Similarly, for the plane-plane joint between parts, we can get the identical results as equations (15) and (16).

5. Illustrative example

5.1 Analysis of the functional requirement

As shown in figure 6, an example of the centring pin assembly consisted of three parts [9] is employed to verify the effectiveness of the proposed analysis line approach. In this example, the ending features of the assembly FR are point P and datum plane A. We wish to control the variation of the vertical displacement of the centring pin endpoint P relative to datum plane A to guarantee the performance of this assembly. According to figure 6, the assembly FR is discretized into point displacement. We need to determine whether the maximum displacement of center pin endpoint P relative to datum A in the vertical direction meets the assembly FR.

IOP Conf. Series: Materials Science and Engineering **626** (2019) 012020 doi:10.1088/1757-899X/626/1/012020

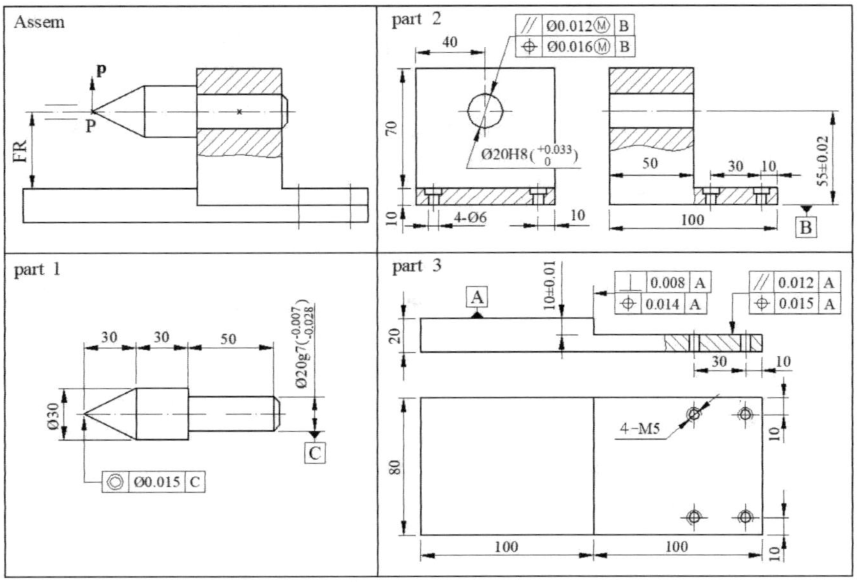

Figure 6. A three parts centring pin mechanism.

5.2 Transfer and accumulation of variations

The displacement of point P in the **p** direction depends on its coaxiality deviation relative to the axis of part1, denoted δ (P^1, **z**) and the variation of point M^1 on the axis of part 1 due to the default of positioning surface (see figure 7), denoted ε(M^1, -**z**), one get:

$$\varepsilon(P, \mathbf{p}) = \delta(P^1, \mathbf{z}) + \varepsilon(M^1, -\mathbf{z}) \tag{17}$$

where δ(P^1, **z**) is determined by the concentricity tolerance, so one gets: $\delta(P^1,\mathbf{z}) = 0.015/2 = 0.0075$. The analysis line of point M^1 on the part1 is along -**z**, the displacement of point M^1 depends on the deviation δ(M^1, -**z**) caused by the clearance and position default between part 1 and part 2 and the deviation of point M^2 on part 2, one gets:

$$\varepsilon(M^1, -\mathbf{z}) = \delta(M^1, -\mathbf{z}) + \varepsilon(M^2, -\mathbf{z}) \tag{18}$$

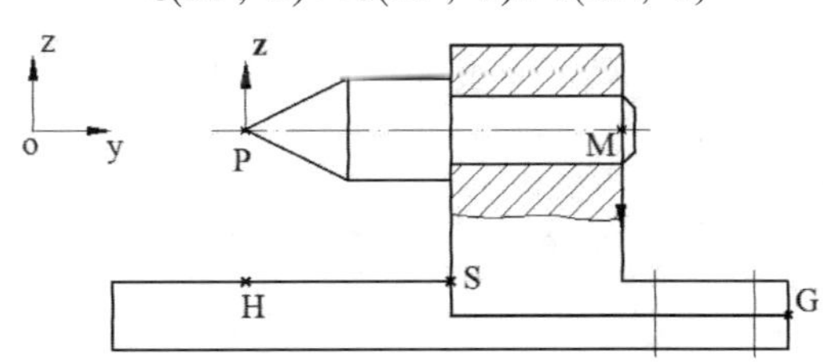

Figure 7. The analysis points of the mechanism

The joint between part 1 and part 2 is a cylinder joint with MMR, the clearance and position default cause the deviation denoted as δ(M^1, -**z**). The maximum material size of the shaft is d_M=19.993 mm; the maximum material virtual size of the hole defined by the position tolerance is D_{po}=19.984 mm; the maximum material virtual size of the hole defined by the parallelism tolerance is D_{pa}=19.988 mm. According to equation (15), one get:

$$\delta(M^1,-\mathbf{z}) = (D_{po}-d_M)/2-(D_{pa}-d_M)\cdot 60/50 = 0.0015 \tag{19}$$

The displacement of point M^2 along -**z** can be expressed as the variation of the hole of part 2 relative to the datum B and the deviation of the analysis point G^2 on part 2.

$$\varepsilon(M^2, -\mathbf{z}) = \delta(M^2, -\mathbf{z}) + \varepsilon(G^2, -\mathbf{z}) \quad (20)$$

where $\delta(M^2, -\mathbf{z})=0.02$. According to the analysis in section 4.3, part 2 is positioned by two planes of part 3, we have:

$$\varepsilon(G^2, -\mathbf{z})=\delta(G^2, -\mathbf{z})+ \delta(G^3, -\mathbf{z})+ \varepsilon(H^3, -\mathbf{z})=0.015/2+0.01+0.014/2+0.008\times 100/10=0.1045 \quad (21)$$

Substituting equations (18)-(21) into equation (17), we obtain the maximum displacement of P is: $\varepsilon(P, \mathbf{p})=0.1335$.

6. Conclusions

Under the assumption of small displacement, the analysis line approach treats the functional tolerance for feature as small displacement variation, and thus converts geometric functional requirement and tolerance variation into small displacements of finite points along the direction of the analysis line. In the worst case, the three-dimensional geometric deviation transfers and accumulation relations of different assembly joints (cylinder-plane joint, plane-plane joint, etc) are studied based on the SDT theory. In order to establish the variation transfer relations of the orthogonal assembly joints, it is necessary to consider the influence of geometric deviations of parts, assembly clearances and the priority order of positioning features between parts. After having analyzed the displacement variations of the typical positioning features, the maximum displacement expression of each point on the functional surface is obtained by using the transfer relation equation, and the three-dimensional tolerance analysis in worst case is realized. This work mainly studies the variation transfer relations and tolerance accumulation calculations of orthogonal assembly joints. The future research will further extend the analysis line approach to complex assembly joints to perform statistical tolerance analysis.

Acknowledgments

The authors would like to acknowledge the financial support by the National Natural Science Foundation of the People's Republic of China (No. 51575235).

References

[1] Marziale M and Polini W 2011 A review of two models for tolerance analysis of an assembly: Jacobian and torsor *Int. J. Comput. Integ. Manuf.* **24** pp 74-86
[2] Desrochers A and Rivière A 1997 A matrix approach to the representation of tolerance zones and clearances *Int. J. Adv. Manuf. Technol.* **13** pp 630-636
[3] Gao J, Chase K W and Magleby S P 1998 Generalized 3-D tolerance analysis of mechanical assemblies with small kinematic adjustments. *IIE. Trans.* **30** pp 367-377
[4] Desrochers A, Ghie W and Laperrière L 2003 Application of a unified Jacobian-Torsor model for tolerance analysis *J. Comput. Inf. Sci. Eng.* **3** pp 2-14
[5] Mujezinović A, Davidson J K and Shah J J 2004 A new mathematical model for geometric tolerances as applied to polygonal faces *J. Mech. Design.* **126** pp 504-518
[6] Anselmetti B 2006 Generation of functional tolerancing based on positioning features. *Comput.-Aided. Des.* **38** pp 902-919
[7] Anselmetti B 2012 ISO manufacturing tolerancing: three-dimensional transfer with analysis line method *Int. J. Adv. Manuf. Technol.* **61** pp 1085-1099
[8] Royer M and Anselmetti B 2016 3D manufacturing tolerancing with analysis line method taking into account joining operations during manufacturing process *Procedia. CIRP.* **43** pp 88-93
[9] Peng H and Wang B 2018 A statistical approach for three-dimensional tolerance redesign of mechanical assemblies *Proc. Inst. Mech. Eng. Part C J. Mech. Eng. Sci.* **232** pp 2132-2144

IOP Conf. Series: Materials Science and Engineering **626** (2019) 012021 doi:10.1088/1757-899X/626/1/012021

Experimental research of nonlinear damage diagnosis using ARMA/GARCH method

Huiyong Guo[1], Rong Zhou[1]

[1] School of civil Engineering, Chongqing University, Chongqing 400045, P. R. China

Email: ghy267@hotmail.com

Abstract. Nonlinear damage such as fatigue crack often occurs in some engineering structures, which brings challenges in structural damage diagnosis. A nonlinear damage diagnosis method based on ARMA/GARCH model is proposed to extract the nonlinear characteristic caused by cracks. First, the background description of ARMA/GARCH model is given and a damage diagnosis index based on this model is proposed to detect nonlinear damage location. The main advantage of ARMA/GARCH model is that this hybrid model can well simulate structural time domain acceleration response and effectively extract the nonlinear damage feature. An experiment of three-storey frame structure, which utilized a mechanism constituted of a bumper and a center column to model breathing crack, is studied. The results show that the damage diagnosis index based on ARMA/GARCH model exhibits good accuracy.
Keywords: damage diagnosis; crack; time series model; acceleration response; nonlinearity

1. Introduction

Structural damage diagnosis has received much attention in recent twenty years. The feasibility of vibration-based damage diagnosis is that the damage usually changes the structural characteristics like stiffness, which are reflected in the measured vibration signals of the structure [1]. Time-domain analysis method is an important damage diagnosis method, which can construct time series model to simulate acceleration response data. Time-domain analysis method has attracted attention in recent years and damage is diagnosed by extracting damage features from the constructed time series models[2].

Time series models can be divided into two categories: linear models and nonlinear models. Linear models include AR model, ARX model and ARMA model etc., which can be utilized to address the linear system problem. Liu et al [3] presented a non-probabilistic damage diagnosis approach based on coefficient matrix of AR model to find diagnosis elements and utilized Mahalanobis distance of the diagnosis elements to identify damage. Damage such as fatigue crack usually possesses time domain nonlinear characteristic caused by opening and closing of crack due to dynamic loading [4]. Nonlinear models include ARCH, GARCH etc., which can be used to address the nonlinear system problem. Pham et al. [5] proposed a hybrid model of ARMA and GARCH to diagnose the machine fault state using measurement vibration signals. Cheng et al. [6] employed a hybrid AR/ARCH model to detect nonlinear damage sites of an eight-storey shear structure and proposed a second-order-variance index. In this paper, a nonlinear damage diagnosis method based on ARMA/GARCH model is proposed to identify the nonlinear characteristic caused by cracks.

2. ARMA/GARCH Model and its application to damage diagnosis

2.1. ARMA/GARCH Model

ARMA models can be utilized to describe structural acceleration time series and evaluate conditional mean. The ARMA (p, q) model for conditional mean can be written as

$$y_t = c + \sum_{i=1}^{p} \varphi_i y_{t-i} + \sum_{j=1}^{q} \theta_j \varepsilon_{t-j} + \varepsilon_t \tag{1}$$

in which y_t is the time series data, c is a constant, p is the number of autoregressive order, φ_i is autoregressive parameter, q is the number of moving average order, θ_j is moving average parameter, ε_t is the residual. Generally, the residual is assumed to be zero mean and constant variance. However, for some practical time domain dynamic responses, the residual does not meet the homoscedastic assumption of constant variance. The time-varying variance is named as conditional variance. The conditional variance σ_t^2 of the residual ε_t is given by

$$\sigma_t^2 = \mathrm{Var}_{t-1}(\varepsilon_t) = \mathrm{E}_{t-1}(\varepsilon_t^2) \tag{2}$$

A GARCH (r, m) model for the conditional variance of residual ε_t is

$$\sigma_t^2 = s + \sum_{i=1}^{r} B_i \sigma_{t-i}^2 + \sum_{j=1}^{m} A_j \varepsilon_{t-j}^2 \tag{3}$$

where s, B_i (i=1,…,r) and A_j (j=1,…,m) are the parameters of GARCH model. The values of these parameters are constrained by

$$s > 0, B_i \geq 0 (i = 1, \cdots, r), A_j \geq 0 (j = 1, \cdots, m) \tag{4}$$

$$\sum_{i=1}^{r} B_i + \sum_{j=1}^{m} A_j < 1 \tag{5}$$

The first three constraints ensure that the conditional variances are positive, and the fourth constraint ensures the covariance stationarity. Thus, the hybrid ARMA/GARCH model can be described as follows:

$$\begin{aligned} & y_t = c + \sum_{i=1}^{p} \varphi_i y_{t-i} + \sum_{j=1}^{q} \theta_j \varepsilon_{t-j} + \varepsilon_t \quad \varepsilon_t \sim N(0, \sigma_t^2) \\ & \sigma_t^2 = s + \sum_{i=1}^{r} B_i \sigma_{t-i}^2 + \sum_{j=1}^{m} A_j \varepsilon_{t-j}^2 \\ & \varepsilon_t = \sigma_t z_t, \ z_t \sim N(0,1) \end{aligned} \tag{6}$$

2.2. Nonlinear damage diagnosis

Here, we consider two states: undamaged state and damaged state. The mainly steps applying the ARMA/GARCH model to calculate nonlinear damage index are summarized as follows:

Firstly, acquire time series acceleration responses under undamaged and damaged states.

Secondly, construct the ARMA models using structural acceleration responses, estimate parameters in the ARMA models and obtain the residual sequences $\{\varepsilon_t^U\}$ and $\{\varepsilon_t^D\}$ under undamaged and damaged states, respectively.

Thirdly, test heteroscedastic effect of the residual sequences $\{\varepsilon_t^U\}$ and $\{\varepsilon_t^D\}$. Then, construct the GARCH models using the residual sequences and obtain the conditional variance sequences $\{\sigma_t^{2^U}\}$ and $\{\sigma_t^{2^D}\}$ of the residual sequences and further obtain the variance $s\{\sigma_t^{2^U}\}$ and $s\{\sigma_t^{2^D}\}$ of the conditional variance sequences under undamaged and damaged states.

At last, establish the damage index based on ARMA/GARCH model. For undamaged structure, the acceleration response of a structure can be approximately regarded as linear data and can be adequately captured by the ARMA model. The residual sequences generated from ARMA model are considered as approximate white noise processes. Thus, the fluctuation of the conditional variances sequence obtained from the GARCH model is close to 0. On the other hand, when there is nonlinear damage occurring, the conditional variances sequences will vary with time. The variance of the conditional variances sequences is much greater than 0. Therefore, for an n-storey structure, the damage index based on ARMA/GARCH model is proposed as follows:

$$c_i = \frac{s_i\left\{\sigma_t^{2^D}\right\} - s_i\left\{\sigma_t^{2^U}\right\}}{\sum_{i=1}^{n} s_i\left\{\sigma_t^{2^D}\right\} - s_i\left\{\sigma_t^{2^U}\right\}} \tag{7}$$

3. Experiment

An experiment of three-storey frame structure was carried out to verify the effectiveness of the proposed method. The frame structure with sensors [7] is shown in Fig. 1. A mechanism constituted of a bumper and a center column is utilized to simulate breathing crack and it will induce nonlinear behaviors when the center column contacts the bumper mounted on the second floor. The distance between the bumper and the center column is adjustable. Thus, the greater the distance is, the smaller the nonlinear damage becomes.

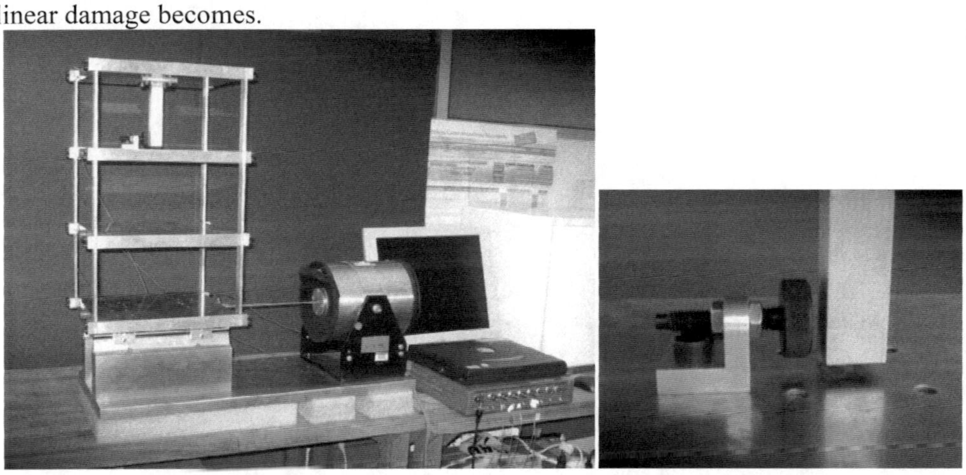

(a) Frame and vibration system (b) Bumper and center column.

Fig. 1. Three-storey frame

Accelerometers are installed on each floor to record time series data. The channels of sensors and parameters are listed in Table 1. The acceleration time series data of the 1st-storey, 2nd-storey and 3rd-storey of the structure are the measurement data of Channels 3, 4 and 5. The damaged scenarios are listed in Table 2. The greater the gap distance is, the smaller the nonlinear damage degree becomes. Therefore, from damage scenarios 1 to 6, the nonlinear damage degree gradually decreases. If the distance is great enough and the center column doesn't contact the bumper in the vibration process, we think that there isn't nonlinear damage behavior. Thus, scenarios 5 and 6 can be regarded as weak nonlinearity problem. The damage scenario 6 considers the disturbance of additional mass, which can simulate the model error. The acceleration responses in undamaged state are depicted in Fig. 2, and the acceleration responses in damaged scenario 1 are depicted in Fig. 3. Akaike information criteria are adopted for order determination of the hybrid model. Here, for the hybrid model, p is 30, q is 0, r is 0, and m is 5.

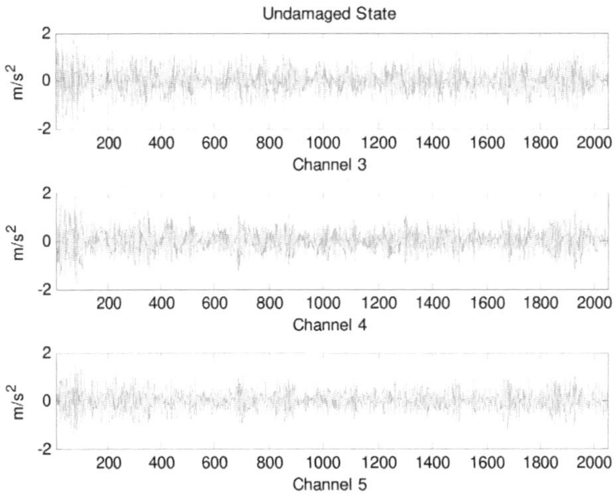

Fig. 2. Acceleration responses in undamaged State.

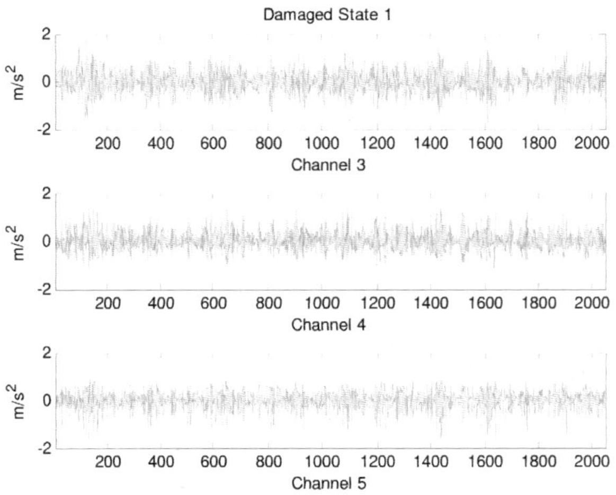

Fig. 3. Acceleration responses in damaged scenario 1.

Table 1. Sensor channels and parameters

Channels	Sensor	Parameter	Sensitivity
Channel 1	Load cell	PCB 208 C03 SN 22569	2.2 mV/N
Channel 2	Accelerometer	PCB 336C SN 10099	1000 mV/g
Channel 3	Accelerometer	PCB 336C SN 10120	1000 mV/g
Channel 4	Accelerometer	PCB 336C SN 9916	1000 mV/g
Channel 5	Accelerometer	PCB 336C SN 10100	1000 mV/g

IOP Conf. Series: Materials Science and Engineering **626** (2019) 012021 doi:10.1088/1757-899X/626/1/012021

Table 2 Damage scenarios of three-storey frame structure

Damaged scenarios	Description
Scenario 1	Gap is 0.05 mm
Scenario 2	Gap is 0.10 mm
Scenario 3	Gap is 0.13 mm
Scenario 4	Gap is 0.15 mm
Scenario 5	Gap is 0.20 mm (weak nonlinearity)
Scenario 6	Gap (0.20 mm) and mass (1.2 kg) at the base (weak nonlinearity)

For damaged scenarios 1-3, acceleration response signals are used to realize nonlinear damage diagnosis. The diagnosis results for damaged scenarios 1-3 are shown in Fig. 4. It can be seen that the calculation values of the 2nd and 3rd storeys are obviously higher than those of the 1st storey, which means that the third inter-storey stiffness next to the 2nd and 3rd storeys is the damage location. The index values of two storeys next to damaged inter-storey stiffness decrease along with the gap distance increasing from scenario 1 to scenario 3 (which means that the degree of nonlinear damage decreases). Therefore, the proposed damage index has the capacity of detecting damage severity.

For damaged scenarios 4-6, the damage index based on ARMA/GARCH model is applied to nonlinear damage extraction. The diagnosis results for damaged scenarios 4-6 are shown in Fig. 5. We can see that the calculation values of the 2nd and 3rd storeys are still higher than those of the 1st storey, which means that the third inter-storey stiffness is the damage location. The index values of the 2nd and 3rd storeys decrease along with the gap distance increasing from scenario 4 to scenario 5, which also implies the method has the potential to quantify the degree of nonlinear damage. In addition, from scenario 6 we can find that the disturbance of additional mass also reduces the diagnosis accuracy.

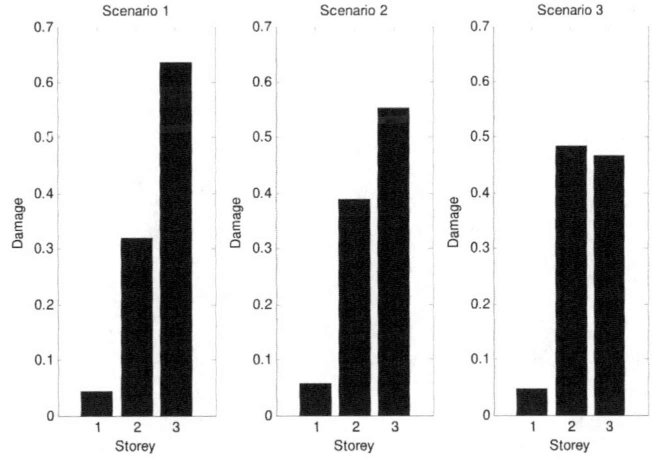

Fig. 4. Nonlinear damage diagnosis results for scenarios 1-3.

IOP Conf. Series: Materials Science and Engineering **626** (2019) 012021 doi:10.1088/1757-899X/626/1/012021

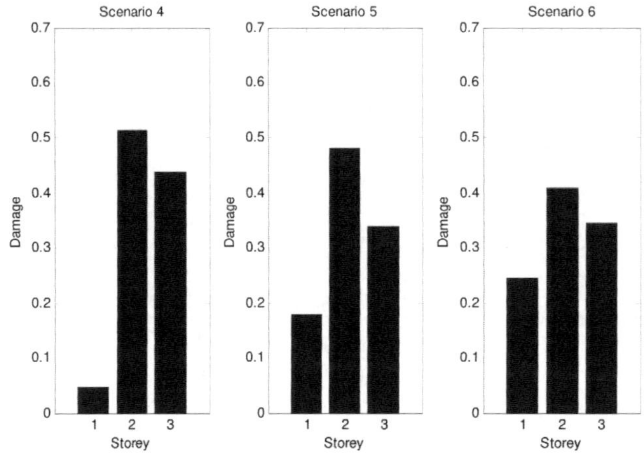

Fig. 5. Nonlinear damage diagnosis results for scenarios 4-6.

4. Conclusions

This paper proposed a damage index based on ARMA/GARCH model to identify nonlinear damage. The basic theories and main procedure are described in this paper and a three-storey frame structure experiment is employed to explore the effectiveness of the proposed method. The results showed that the damage index has good performance to locate the nonlinear damage and has the potential to reflect the changed degree of nonlinear damage caused by adjustable gap distance. In addition, due to the fact that the damage index is only rely on the output acceleration response data, it has the possibility to combine with continuous online monitoring to monitor the health condition of structure in real time.

Acknowledgments

This work was supported by the National Natural Science Foundation of China (Grant No. 51578094).

5. References

[1] Li P J, Xu D W and Zhang J. 2016 Probability-based structural health monitoring through Markov chain Monte Carlo sampling. *International Journal of Structural Stability & Dynamics*, **16**, 1550039-1-16.

[2] Musafere F, Sadhu A and Liu K. 2016 Towards damage detection using blind source separation integrated with time-varying auto-regressive modeling. *Smart Materials and Structures*, **25**(1), 1-19.

[3] Liu G, Mao Z and Luo J. 2015 Damage detection with interval analysis for uncertainties quantification. *International Conference on Performance-based and Life-cycle Structural Engineering*, Queensland, Australia, 10 January, p.272-281

[4] Yu L and Zhu J H 2015 Nonlinear damage detection using higher statistical moments of structural responses. *Structural Engineering & Mechanics* **54**, 221-237.

[5] Pham H T and Yang B S. 2010 Estimation and forecasting of machine health condition using ARMA/GARCH model. *Mechanical Systems and Signal Processing*, **24**, 546-558.

[6] Cheng J J, Guo H Y, Wang Y S. 2017 Structural Nonlinear Damage Detection Method Using AR/ARCH Model. *International Journal of Structural Stability and Dynamics*, **17**, 1750083-1-27.

[7] Figueiredo E, Park G, Figueiras J et al. 2009 Structural health monitoring algorithm comparisons using standard data sets. *Los Alamos National Laboratory* (LANL), Los Alamos, NM, United States.

IOP Conf. Series: Materials Science and Engineering **626** (2019) 012022 doi:10.1088/1757-899X/626/1/012022

Prediction of Collapse Cycles of Shale Wellbores in Bonger Basin of Chad

Zhang yanna[1 *], Luo huaidong[2], Shi Libao[1]

[1]PetroChina Research Institute of Petroleum Exploration & Development, China
[2]CNPC international(Chad)Co., Ltd., China

zh-yanna@qq.com

Abstract. This paper presents a new collapse cycle model, which is based on the distribution of stress surrounding boreholes under mechanical action and the distribution of pore pressure under chemical action. By considering rock failure criteria, the model has been established for estimating the collapse cycle of shale wellbores. Equation parameters have been verified by experimental studies of Rock mechanics and drilling fluid. The model accounts for the effect of inclination angle and density of drilling fluid on the collapse cycle. The model has been used in 22 directional wells drilled in Chad. According to simulation calculation, the collapse cycle of mudstone wellbores of drilled wells is about 8 d, which is approximately equal to the drilling cycle. The wellbores are stable during drilling process, but they are easy to collapse during well completion, which is consistent with actual drilling data.

1. Introduction

The Bonger Basin in Chad is located in the intersection of the middle and western African valleys, with complex structures and stresses. The Lower Cretaceous formation developed shale, which is easy to spall and disintegrate to result in over pull or stick during drilling. These issues have seriously restricted the progress and quality of safety drilling operations, and become bottlenecks in the development of oilfields in Chad. In this paper, the model has been established for calculating the wellbore collapse cycle. The model has been used to accurately predict the collapse cycle of shale wellbores and can provide guidance for field drilling operations.

2. Formatting the title, authors and affiliations

To effectively improve the stability of shale wellbores, it is necessary to provide the collapse cycle of shale wellbores under water-base drilling fluid and to guide field operations. In this paper, a stress distribution model and a pore pressure distribution model have been established near wellbore in Chad; and based on which the wellbore collapse cycle can be estimated by combining applicable wellbore failure criteria.

2.1. Stress distribution model

The stresses ( ) are consistent with the principal axes of the coordinate system (h, H, v). The deviation of the borehole axis is described by azimuth (dw) and well inclination angle (iw). The well inclination angle is the angle between the borehole axis and the vertical axis (the axis of overburden formation pressure). The azimuth is the angle between the projection of the borehole axis on plane and the horizontal principal stress. When the borehole axis is consistent with Z direction, generally Cartesian coordinate system (xyz) and cylindrical coordinate system (r θ z) are used to describe the stress field near the wellbore.

Converted from a (h, H, v) coordinate system to a (xyz) coordinate system, the second-order stress tensor σ_{ij} expressed by $\sigma_h, \sigma_H, \sigma_v$ is:

$$\begin{cases} \sigma_x = cos^2\varphi cos^2\alpha\sigma_H + \sigma_H sin^2\varphi cos^2\alpha + \sigma_v sin^2\alpha \\ \sigma_y = \sigma_H sin^2\varphi + \sigma_h cos^2\varphi \\ \sigma_z = cos^2\varphi sin^2\alpha\sigma_H + \sigma_h sin^2\varphi sin^2\alpha + \sigma_v cos^2\alpha \\ \tau_{xy} = \frac{1}{2}(\sigma_h - \sigma_H) cos\alpha sin^2\varphi \\ \tau_{yz} = \frac{1}{2}(\sigma_h - \sigma_H) sin2\varphi sin\alpha \\ \tau_{xz} = \frac{1}{2}(\sigma_H cos^2\varphi + \sigma_h sin^2\varphi - \sigma_v) sin2\alpha \end{cases} \tag{1}$$

In a cylindrical coordinate system, the total stress caused by in-situ stress is the linear superposition of the stress caused by various in-situ stresses.

The stress caused by is:

$$\begin{cases} \sigma_{rr} = \left(1 - \frac{r_w^2}{r^2}\right)\frac{\sigma_{xx}}{2} + \left(1 - 4\frac{r_w^2}{r^2} + 3\frac{r_w^4}{r^4}\right)\frac{\sigma_{xx}}{2}\cos 2\theta \\ \sigma_{\theta\theta} = \left(1 + \frac{r_w^2}{r^2}\right)\frac{\sigma_{xx}}{2} + \left(1 + 3\frac{r_w^4}{r^4}\right)\frac{\sigma_{xx}}{2}\cos 2\theta \\ \tau_{r\theta} = -\left(1 + 2\frac{r_w^2}{r^2} - 3\frac{r_w^4}{r^4}\right)\frac{\sigma_{xx}}{2}\sin 2\theta \end{cases} \tag{2}$$

The stress caused by σ_{yy} is:

$$\begin{cases} \sigma_{rr} = \left(1 - \frac{r_w^2}{r^2}\right)\frac{\sigma_{yy}}{2} - \left(1 - 4\frac{r_w^2}{r^2} + 3\frac{r_w^4}{r^4}\right)\frac{\sigma_{yy}}{2}\cos 2\theta \\ \sigma_{\theta\theta} = \left(1 + \frac{r_w^2}{r^2}\right)\frac{\sigma_{yy}}{2} + \left(1 + 3\frac{r_w^4}{r^4}\right)\frac{\sigma_{yy}}{2}\cos 2\theta \\ \tau_{r\theta} = \left(1 + 2\frac{r_w^2}{r^2} - 3\frac{r_w^4}{r^4}\right)\frac{\sigma_{yy}}{2}\sin 2\theta \end{cases} \tag{3}$$

σ_{zz} as:

$$\sigma_{zz} = \sigma_z - v\left(2(\sigma_{xx} - \sigma_{yy})\frac{r_w^2}{r^2}\cos 2\theta + 4\sigma_{xy}\frac{r_w^2}{r^2}\sin 2\theta\right) \tag{4}$$

The stress caused by σ_{xy} is:

$$\begin{cases} \sigma_{rr} = \left(1 - 4\frac{r_w^2}{r^2} + 3\frac{r_w^4}{r^4}\right)\sigma_{xy}\sin 2\theta \\ \sigma_{\theta\theta} = -\left(1 + 3\frac{r_w^4}{r^4}\right)\sigma_{xy}\sin 2\theta \\ \tau_{r\theta} = \left(1 + 2\frac{r_w^2}{r^2} - 3\frac{r_w^4}{r^4}\right)\sigma_{xy}\cos 2\theta \end{cases} \tag{5}$$

The stress caused by σ_{xz} is:

$$\begin{cases}\tau_{rz} = \sigma_{xz}\left(1-\frac{r_w^2}{r^2}\right)\cos\theta \\ \tau_{\theta z} = -\sigma_{xz}\left(1+\frac{r_w^2}{r^2}\right)\sin\theta\end{cases} \tag{6}$$

The stress caused by σ_{yz} is:

$$\begin{cases}\tau_{rz} = \sigma_{yz}\left(1-\frac{r_w^2}{r^2}\right)\sin\theta \\ \tau_{\theta z} = -\sigma_{yz}\left(1+\frac{r_w^2}{r^2}\right)\cos\theta\end{cases} \tag{7}$$

In summary, the stress distribution near the wellbore caused by in-situ stress is:

$$\sigma_{rr} = \left(1-\frac{r_w^2}{r^2}\right)\frac{\sigma_{xx}}{2} + \left(1-4\frac{r_w^2}{r^2}+3\frac{r_w^4}{r^4}\right)\frac{\sigma_{xx}}{2}\cos 2\theta + \left(1-\frac{r_w^2}{r^2}\right)\frac{\sigma_{yy}}{2} - \left(1-4\frac{r_w^2}{r^2}+3\frac{r_w^4}{r^4}\right)\frac{\sigma_{yy}}{2}\cos 2\theta + \left(1-4\frac{r_w^2}{r^2}+3\frac{r_w^4}{r^4}\right)\sigma_{xy}\sin 2\theta$$

$$\sigma_{\theta\theta} = \left(1+\frac{r_w^2}{r^2}\right)\frac{\sigma_{xx}}{2} + \left(1+3\frac{r_w^4}{r^4}\right)\frac{\sigma_{xx}}{2}\cos 2\theta + \left(1+\frac{r_w^2}{r^2}\right)\frac{\sigma_{yy}}{2} + \left(1+3\frac{r_w^4}{r^4}\right)\frac{\sigma_{yy}}{2}\cos 2\theta - \left(1+3\frac{r_w^4}{r^4}\right)\sigma_{xy}\sin 2\theta$$

$$\sigma_{zz} = \sigma_z - v\left(2(\sigma_{xx}-\sigma_{yy})\frac{r_w^2}{r^2}\cos 2\theta + 4\sigma_{xy}\frac{r_w^2}{r^2}\sin 2\theta\right)$$

$$\tau_{r\theta} = -\left(1+2\frac{r_w^2}{r^2}-3\frac{r_w^4}{r^4}\right)\frac{\sigma_{xx}}{2}\sin 2\theta + \left(1+2\frac{r_w^2}{r^2}-3\frac{r_w^4}{r^4}\right)\frac{\sigma_{yy}}{2}\sin 2\theta + \left(1+2\frac{r_w^2}{r^2}-3\frac{r_w^4}{r^4}\right)\sigma_{xy}\cos 2\theta$$

$$\tau_{rz} = \sigma_{xz}\left(1-\frac{r_w^2}{r^2}\right)\cos\theta + \sigma_{yz}\left(1-\frac{r_w^2}{r^2}\right)\sin\theta$$

$$\tau_{\theta z} = -\sigma_{xz}\left(1+\frac{r_w^2}{r^2}\right)\sin\theta - \sigma_{yz}\left(1+\frac{r_w^2}{r^2}\right)\cos\theta \tag{8}$$

2.2. *Pore pressure distribution model*

Hydraulic pressure difference and chemical potential difference are two main factors that cause the change of pore pressure, thus the volume flow rate through a cell cube can be expressed as:

$$J = \frac{k}{\mu}\frac{1}{\Delta x}\left(\Delta p - I_m\left(\frac{RT}{v}\ln\left(\frac{a_s}{a_d}\right)\right)\right) \tag{9}$$

where, J=volume flow rate, mL/cm3; k=permeability of rock, μm^2; μ =fluid viscosity, $mPa \cdot s$.
It can be obtained from balance equation (in a cylindrical coordinate):

$$\frac{1}{r}\frac{\partial}{\partial r}(r\rho J) + \frac{\partial}{\partial}(\varphi\rho) = 0 \tag{10}$$

It can be obtained from equation (9) and equation (10):

$$\frac{1}{r}\frac{\partial p}{\partial r} + C\left(\frac{\partial p}{\partial r}\right)^2 + \frac{\partial^2 p}{\partial r^2} + \frac{1}{r}I_m RT\frac{\partial a}{\partial r} + I_m\frac{RT}{v}C\frac{\partial a}{\partial r} + I_m\frac{RT}{v}\frac{\partial^2 a}{\partial r^2} = \frac{\varphi\mu C}{k}\frac{\partial p}{\partial t} \tag{11}$$

where, C=fluid compressibility; φ =porosity of rock.
Activity degree a should satisfy the diffusion equation:

$$\frac{\partial a}{\partial t} - D\nabla^2 a = 0 \tag{12}$$

where, D=diffusion coefficient of activity degree.
The boundary condition and initial condition are:

when t=0, $r_w \le r \le \infty$, $a = a_s$, $p = p_o$;

when t>0, $r = r_w$, $a = a_d$, $p = p_w$;

when t<0, $r = \infty$, $a = a_s$, $p = p_o$.

where, r=the distance from wellbore axis, m; r_w =wellbore radius, m; p_w =bottom hole pressure, MPa. By combining boundary conditions and equation (11), the distribution law of pore pressure is obtained.

2.3. Total stress distribution

The total stress distribution near wellbore under the effects of hydraulic, chemical and stress is:

$$\begin{aligned}\sigma_{rr} &= \left(1-\frac{r_w^2}{r^2}\right)\frac{\sigma_{xx}}{2} + \left(1-4\frac{r_w^2}{r^2}+3\frac{r_w^4}{r^4}\right)\frac{\sigma_{xx}}{2}\cos 2\theta + \left(1-\frac{r_w^2}{r^2}\right)\frac{\sigma_{yy}}{2} \\ &- \left(1-4\frac{r_w^2}{r^2}+3\frac{r_w^4}{r^4}\right)\frac{\sigma_{yy}}{2}\cos 2\theta + \left(1-4\frac{r_w^2}{r^2}+3\frac{r_w^4}{r^4}\right)\sigma_{xy}\sin 2\theta \\ &+ \frac{\alpha(1-2v)}{(1-v)}\frac{1}{r^2}\int_{r_w}^{r} p^f(r,t)r\,dr\end{aligned} \tag{13}$$

$$\begin{aligned}\sigma_{\theta\theta} &= \left(1+\frac{r_w^2}{r^2}\right)\frac{\sigma_{xx}}{2} + \left(1+3\frac{r_w^4}{r^4}\right)\frac{\sigma_{xx}}{2}\cos 2\theta + \left(1+\frac{r_w^2}{r^2}\right)\frac{\sigma_{yy}}{2} \\ &+ \left(1+3\frac{r_w^4}{r^4}\right)\frac{\sigma_{yy}}{2}\cos 2\theta - \left(1+3\frac{r_w^4}{r^4}\right)\sigma_{xy}\sin 2\theta \\ &- \frac{\alpha(1-2v)}{(1-v)}\left[\frac{1}{r^2}\int_{r_w}^{r} p^f(r,t)r\,dr - p^f(r,t)\right]\end{aligned} \tag{14}$$

$$\begin{aligned}\sigma_{zz} &= \sigma_z - v\left(2(\sigma_{xx}-\sigma_{yy})\frac{r_w^2}{r^2}\cos 2\theta + 4\sigma_{xy}\frac{r_w^2}{r^2}\sin 2\theta\right) \\ &+ \frac{v\alpha(1-2v)}{(1-v)}p^f(r,t)\end{aligned} \tag{15}$$

$$\begin{aligned}\sigma_{r\theta} &= -\left(1+2\frac{r_w^2}{r^2}-3\frac{r_w^4}{r^4}\right)\frac{\sigma_{xx}}{2}\sin 2\theta + \left(1+2\frac{r_w^2}{r^2}-3\frac{r_w^4}{r^4}\right)\frac{\sigma_{yy}}{2}\sin 2\theta \\ &+ \left(1+2\frac{r_w^2}{r^2}-3\frac{r_w^4}{r^4}\right)\sigma_{xy}\cos 2\theta\end{aligned} \tag{16}$$

$$\sigma_{\theta z} = \sigma_{yz}\left(1+\frac{r_w^2}{r^2}\right)\cos\theta - \sigma_{xz}\left(1+\frac{r_w^2}{r^2}\right)\sin\theta \tag{17}$$

$$\sigma_{rz} = \sigma_{xz}\left(1-\frac{r_w^2}{r^2}\right)\cos\theta + \sigma_{yz}\left(1-\frac{r_w^2}{r^2}\right)\sin\theta \tag{18}$$

2.4. Rock failure criteria

Rock failure criteria are critical conditions for judging whether a wellbore is damaged. In the study, the Mohr-Coulomb criterion was selected for shear failure. By equation derivation, the stress distribution can be obtained within a certain scope near wellbore; by combining the Mohr-Coulomb rock failure criterion, the collapse pressure can be obtained near wellbore under various circumstances. The expression is:

$$(\sigma_1 - \alpha P_p)[(f^2+1)-f] - (\sigma_3 - \alpha P_p)[(f^2+1)+f] = 2\tau_o \quad (19)$$

where, σ_1=the maximum principal stress on a point on the wellbore; σ_3 = the minimum principal stress on a point on the wellbore; α=effective stress factor; P_p=formation pore pressure; τ_o=cohesive force; f=internal friction coefficient.

3. Numerical solution and analysis

Based on field data, well logging data and laboratory experiments, basic parameters were obtained for calculating the collapse cycle of shale wellbores in Chad (Table 1).

Table 1. Formation and petrological parameters

Well depth	1300 m	Well inclination angle	0° ,30° ,60° ,90°
Biot factor	0.85	Azimuth	0°
Poisson ratio	0.234	Vertical stress	2.22 g/cm3
Elastic modulus	3.40 GPa	Maximum horizontal principal stress	1.91 g/cm3
Cohesive force	5.08 MPa	Minimum horizontal principal stress	1.59 g/cm3
Internal friction angle	26.97°	Fluid compressibility	0.8 GPa-1
Porosity	0.0335	Permeability	2 mD
Equivalent density of drilling fluid	1.5 g/cm3	Membrane efficiency	0.3
Viscosity of drilling fluid	5 mPa.s	Activity degree	0.9

The collapse pressure changes with time as follows (Figure 1).

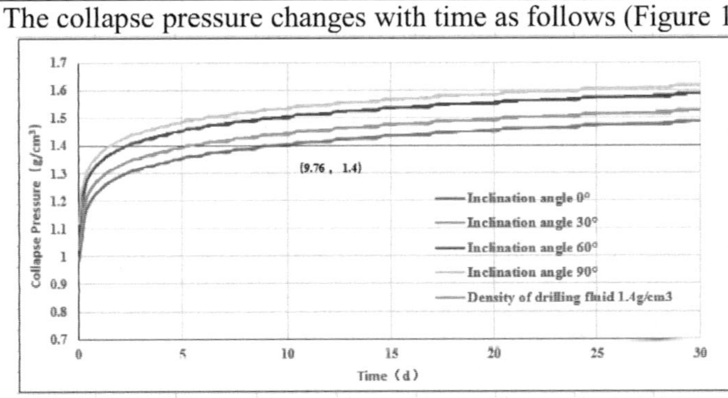

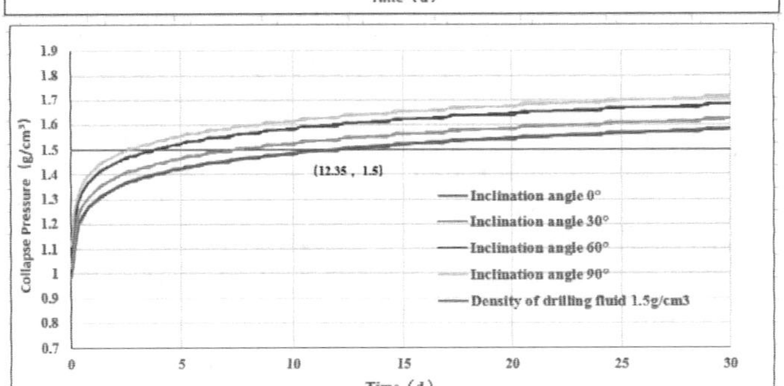

Figure 1 Collapse pressure vs. time at 0 azimuth

Figure1 shows that when azimuth is 0, the collapse pressure increases with the increasing inclination angle; when the density of drilling fluid is 1.5 g/cm3 and the inclination angle is 0°, the longest collapse cycle is 12.35 d.

IOP Conf. Series: Materials Science and Engineering **626** (2019) 012022 doi:10.1088/1757-899X/626/1/012022

4. Conclusions

(1) The lower Cretaceous in the Bonger Basin in Chad develops thick shales which has high clay mineral content and is easy to disintegrate with hydration. Moreover, with well-developed fractures, drilling fluid filtrate can easily invade formations to reduce rock strength and wellbore stability.
(2) Based the stress distribution model and pore pressure model near wellbore, and combining the Mohr-Coulomb rock failure criterion, the model for calculating the collapse cycle of shale wellbores has been obtained; and based on the model, how collapse pressure changes with time is obtained at different well inclination angles and different densities of drilling fluid.
(3) Based on the statistical drilling cycles and drilling incidents of drilled wells, it has been confirmed the consistence of simulated with actual results, and that the model for calculating collapse cycle and the parameters selected are more reasonable. The research results can accurately predict the collapse cycle of shale wellbores, and provide guidance for designing drilling fluid formula and drilling operations in Chad.

References

[1] Wang Bingyin, Deng Jingen, Zou Lingzhan, et al. Applied research of collapse cycle of shale in wellbore using a coupled physico-chemical model[J]. Acta Petrolei Sinica, 2006, 27(3):130-132.

[2] Deng Jingen, Guo Dongxu, Zhou Jianliang, et al. Mechanics-chemistry coupling calculation model of boreholestressing in shale formation and its numerical solving method[J]. Chinese Journal of rock Mechanics and Engineering, 2003, 22(s1):2250-2253.

[3] Cheng Yuanfang, Zhang Feng, Wang Jingyin, et al. Analysis of borehole collapse cycling time for shale[J]. Journal of China University of Petroleum (Edition of Natural Science), 2007, 31(1): 63-66, 71.

[4] Qu Lianzhong, Cheng Yuanfang, Zhao Yizhong, et al. Fluid solid coupling simulation of deliverability ofscreen out fracturing in permeability anisotropy unconsolidated sandstone[J]. Journal of China University of Petroleum(Edition of Natural Science), 2009, 33(2): 70-74.

[5] Cheng Yuanfang, Li Lingdong, MAHMOOD S, et al. Fluid-solid coupling model for studying wellbore instability in drilling of gas hydrate bearing sediments[J]. Applied Mathematics and Mechanics, 2013, 34(11): 1421-1432.

[6] R.Freij-Ayoub, C.P. Tan. Simulation of Time-Dependent Wellbore Stability in Shales Using A Coupled Mechanical-Thermal-Physico-Chemical Model[J]. SPE 85344, 2003.

[7] M.Yu, G.Chen, M.E.Chenevert. Chemical and Thermal Effects on Wellbore Stability of Shale. SPE71366, 2001.

[8] Liu Xiangjun, Zeng wei, Liang Lixi, et al. Prediction of collapse cycle of shale formation in Longma stream formation[J]. Special oil and gas reservoirs. 2016, 23(5): 1-8.

[9] Yuan Huayu, Cheng Yuanfang, Wang Wei, et al. Analysis on time-dependent wellbore collapse for long horizontal well in shale formation[J]. Science Technology and Engineering, 2017, 17(3): 183-189.

[10] Yan Chuanliang, Deng Jingen, Yu Baohua, et al. Wellbore stability analysis and its application in the Fergana basin, central Asia[J]. Journal of Geophysics and Engineering, 2014, 11(1):1-9.

[11] Chen G, Ewy R.T., Yu M.Analytic solutions with ionic flow for a pressure transmission test on shale.Journal of Petroleum Science and Engineering[J]. 2010, 72(1):158-165.

[12] Li Yufei, Fu Yongqiang, Tang Geng, et al. Laws of the effects of earth stress patterns on wellbore stability in a directional well. Gas Industry, 2012, 32(3):78-80.